# DISEASES OF VEGETABE CROPS AND THEIR INTEGRATED MANAGEMENT

## — A Colour Handbook —

AF631026

NIPA® GENX ELECTRONIC RESOURCES & SOLUTIONS P. LTD.
New Delhi-110 034

www.nipaers.com
Browse, Search, Read & Buy...
Print Books
eChapters
eBooks
Publishers
Forthcoming
Subject
Catalogues
New Titles
Online Resources on:
✓ Current Affairs
✓ Reasoning, Logic & Aptitude
✓ Competitive Examinations
✓ English Language Lab
✓ Writing & Pronunciation Tools
✓ Personality Development
✓ Online Programmes for Professionals
✓ Interview Preparation
Pay Using
PhonePe
Google Pay
Paytm
PayPal
UPI
CCAvenue

# DISEASES OF VEGETABLE CROPS AND THEIR INTEGRATED MANAGEMENT

## —A Colour Hand Book—

R.K. Mishra
Dr. A.K. Singh
A.K. Sharma

NIPA® GENX ELECTRONIC RESOURCES & SOLUTIONS P. LTD.
New Delhi-110 034

NIPA® GENX ELECTRONIC
RESOURCES & SOLUTIONS P. LTD.

101,103, Vikas Surya Plaza, CU Block
L.S.C. Market, Pitam Pura, New Delhi-110 034
Ph : +91 11 27341616, 27341717, 27341718
E-mail: newindiapublishingagency@gmail.com
www: www.nipabooks.com

***For customer assistance, please contact***
Phone: + 91-11-27 34 17 17
Fax: + 91-11- 27 34 16 16
E-Mail: feedbacks@nipabooks.com



ISBN: 978-81-19072-94-1



Composed and Designed by NIPA®.

**Gurbachan Singh**
Chairman

AGRICULTURAL SCIENTISTS RECRUITMENT BOARD
(INDIAN COUNCIL OF AGRICULTURAL RESEARCH)
Krishi Anusandhan Bhavan-I, Pusa, New Delhi – 110012
Telephone: (0) 25843295, 25841272; Telefax: 25846540
Telegram: AGRECBOARD Email: gurbachansingh@icar.org.in

# Foreword

Vegetable diseases are important factors that are responsible for causing huge loss in vegetable production. The losses due to diseases alone are reported to be around 26 per cent. Diseases may attack at any stage of the standing crop, from nursery till maturity and under storage stages of the crop. They may affect different parts of the plants, such as leaves, stem, roots and seed and cause various types of symptoms. Further, many pathogens continue to attack the stored produce/vegetables and causing spoilage. To save the vegetable crops from diseases caused by pathogens and thereby to increase production, it is imminent that diseases have to be kept below threshhold level to minimize the losses. Chemical methods for diseases and pests management are commonly adopted in many crops the world wide. However, the indiscriminate use has resulted in development of resistance in pests, pest outbreaks/pest on one hand and health hazards to on the other hand. The importance of enpoying sustainable vegetable production technology in through the use of eco-friendly integrated approaches is being realized more and more in the recent past. Thus, with the increasing importance of sustainable vegetable production, the concept of integrated pest management (IPM) has been realized as an important tool for btic stress management. There is an urgent need to document the total scenario of sustainable disease and pest management to develop better plant protection programmes and strategies for food and nutritional security in the near future. The instant book is an effort in that direction.

In this book, the authors have given a detailed account in written & ... forms of the major diseases of important vegetable crops, and their management strategies through IPM approaches.

I not only believe book entitled "Diseases of Vegetable Crops and Their Integrated Management —A Colour Handbook" will be useful to scientists, teachers, extension workers, vegetable growers, Vegetable and Producers students and to those working in the field of vegetables research and development. I congratulate the authors for their effort in bringing out this useful book.

**(Gurbachan Singh)**

# Preface

Over exploitation of natural resources and excessive chemicalization of agriculture in the past few decades, have led to poor sustainability of vegetable production. Over dependency and indiscriminate use of chemicals and fertilizers have resulted in the problems of pest/diseases resurgence, secondary pest out breaks, development of pest resistance, decimation of natural enemies on one hand and disturbances to environment and health hazards through residues in food chain on the other hand. These side effects have raised public concern about the safety of use of pesticides. Integrated Pests Management (IPM) is one such environment friendly approach to pest management that relies on combination of common sense approaches and fits in to the concept of sustainable agriculture. IPM incorporates ecological and socio-economic factors in to agricultural ecosystem and decision making process and addresses public concern about environmental quality and food safety. Today IPM is considered to be the spring board to sustainable management of vegetable crops and form a focus of the present book. Diseases are the most limiting biotic factors in vegetable production. There has been a demand from progressive vegetable growers, plant protection officials, subject matter specialist and research students for long for a comprehensive colour compendium on vegetable diseases. In the absence of such a type of book, identification is still difficult by the personnel engaged in different activities of vegetable production. With this view, **"Diseases of vegetable crops and their integrated management"** is prepared. This book has been prepared to assist clear diagnosis of the diseases and planning of integrated management strategies for sustainability of vegetable production. On the basis of several experiments conducted on vegetable crops for last several years and experience gained on vegetable pathology, integrated management practices have been encapsulated in this book. We hope that this book will serve all needy personnel whose ever is directly or indirectly related to vegetable production.

**R. K. Mishra**
**A. K. Singh**
**A. K. Sharma**

# Glossary

**Aecium:** A cup shaped fruiting body of the rust fungi consisting of binucleate hyphal cells, with or without a peridium that produces spore chains consisting of aeciospores, alternating with disjuncture cells, following successive conjugate division of the nuclei (stage I of the rust fungi)

**Abiotic:** Caused by a non-living agent, for example, stress as like, heat, water, drought etc.

**Aerobic:** A microorganism that lives or a process that occurs in the presence of molecular oxygen.

**Aflatoxin:** Mycotoxins produced by the fungi belonging to genus *Aspergillus*, which are carcinogenic (substance or agent that causes cancer) and sometimes may even be lethal andfatal. They are the derivatives of coumarine.

**Alternate host:** One of two kinds of plants on which a parasitic fungus (e.g. rust) must develop to complete its life cycle.

**Anaerobic:** A microorganism that lives or a process that occurs in the absence of molecular oxygen

**Anamorph:** The imperfect or asexual stage of a fungus

**Antheidium:** The male sexual organ found in some fungi.

**Amphitrichous:** Single flagellum at both the polar ends

**Antheridium:** The male gametangium of heterogamous fungi

**Antibiotic:** A chemical compound produced by a micro organism, which inhibits or kills some other specific microorganisms.

**Apothecium:** An open cup or saucer shaped ascocarp

**Appressorium:** A swollen, flattened, pressing organ arising from the tip of the hyphae or germ tube that facilitates attachment to the host. A minute infection peg usually grows from it and enters the epidermal cell of the host.

**Ascocarp or Ascoma:** Sexual fruiting body of an ascomycetous fungus that produces asci and ascospores

**Ascogonium:** The female gametangium of the ascomycetes

**Ascus:** A sac like containing definite number of ascospores, usually eight, formed by free cell formation, mostly after karyogamy and meiosis

**Autoecious:** A parasitic fungus, which can complete its entire life cycle on a single host species as in the case of some rusts.

**Avirulent:** Lacking virulence, unable to cause disease

**Basidiocarp:** A sexual fruiting body of the basidiomycetous fungi, in or on which basidia and basidiospores are produced

**Basidium:** A cup shaped, zygote cell bearing a definite number of basidiospores, usually four, at the end of minute sterigmata formed at the apex of the basidium, following karyogamy and meiosis.

**Biolobical control:** Manipulation of natural enemies like parasitoides, predators or other microorganisms in an attempt at reducing the pest population and keep them at much reduced levels.

**Biotic agents:** A living organisms that is the cause of damage, ability or disease

**Biotroph:** An organism that can live and multiply only on another living organism

**Blight:** A disease characterized by general and rapid killing of tissues of leaves, flowers and stems.

**Blotch:** A disease characterized by large and irregularly shaped, necrotic spots or blots on leaves, shoots and stems.

**Broad spectrum pesticides**: Pesticides effectives against a variety of pests and diseases.

**Canker:** A necrotic, often sunken or cracked lesion surrounded by callus on a stem, branch or twig of a plant.

**Chemotherapy:** Management of plant diseases with chemotherapeutants, such as systemic fungicides and antibiotics that are absorbed and are translocated internally.

**Chlorosis:** Yellowing of normally green tissues due to chlorophyll destruction or failure of chlorophyll formation

**Chronic**: Slow developing, persistent or recurring symptoms

**Clavate**: Club-like, narrowing in the direction of the base

**Cleistothecium**: A completely closed asocarp containing asci and ascospores.

**Chlamydospoe:** A thick-walled asexual spore formed by the modification of a cell of a fungus hypha.

**Cleistothecium:** An entirely closed ascocarp

**Conjugation:** A process of sexual reproduction involving the fusion of two gametes. As in case of bacteria the transfer of genetic material from a donar cell to a recipient cell through direct cell to cell contact takes place.

**Conidiophore**: A specialized hypha on which one or more conidia are produced.

**Conidium:** A sexual fungus spore formed from the end of conidiophores.

**Cross Protection:** The phenomenon in which plant tissues infected with one strain of a virus are protected from infection by other more severe strains of the same virus.

**Dieback**: Progressive death of shoots, branches and roots generally starting at the tip.

**Dikaryotic:** Mycelium or spore containing two sexually compatible nuclei per cell. Common in case of fungi belonging to basidiomycetes.

**Disease:** Any malfunctioning of host cells and tissues that result from continuous irritation by a pathogenic agent or environment factor and leads to development of symptoms.

**Disease cycle:** The chain of events involved in disease development, including the stages of development of the pathogens and the effect of the disease on the host.

**Disinfectant:** A physical or chemical agent that frees a plant, organ, or tissue from infection.

**Disinfestant:** An agent that kills or inactivates pathogens in the environment or on the surface of a plant or plant organ before infection takes place.

**ELISA (Enzyme Linked Immuno serological Assay):** Serological test in which one antibody carries with it an enzyme that releases a coloured compound.

**Elicitors:** Molecules produced by the pathogen that induce by certain virus infection.

**Epidemic:** A disease increase in a population. Usually it is a widespread and severe outbreak of a disease.

**Epidemic rate**: The amount of increase of disease per unit of time in a plant population.

**Epidemiology:** The study of factors affecting the outbreak and spread of infectious disease.

**Epiphytotic:** A widespread and destructive outbreak of a disease of plant.

**Eradicant:** A chemical substance that destroys a pathogen at its source.

**Erdication:** Control of plant disease by eliminating the pathogen after it is established or by eliminating the plants that carry the pathogen.

**Etiology of disease**: The determination and study of the cause of a disease.

**Facultative parasite:** Having the ability to be a parasite.

**Free-living:** A microorganisms that lives freely, an attached, or pathogen living in the soil, outside its host.

**Fruiting body:** A complex fungal structure containing spores.

**Fungistatic:** A compounds that prevents fungus growth without killing the fungus.

**Gall:** A swelling or overgrowth produced on a plant as a result of infection by certain pathogens.

**Germtube:** The early growth of mycelium produced by a germinating fungus spore.

**Haustorium:** A single or branched projection of hyphae in to host cells which acts as an absorbing organisms.

**Hetroecious:** A fungus requiring two different kinds of hosts to complete its life cycle. It Pertains particularly to rust fungi.

**Heterokaryosis:** The condition in which a mycelium contains two genetically different nuclei per cell.

**Hetrothallic fungi:** Fungi producing compatible male and female gametes on physiologically distinct mycelia.

**Homothallic fungus:** A fungus producing compatible male and female gametes on the same mycelium.

**Horizontal resistance**: Partial resistance equally effective against all races of a pathogen.

**Host**: A plant that is invaded by a parasite and from which the parasite obtains its nutrients.

**Host range:** The various kinds of host plants that may be attacked by a parasite.

**Hybrids**: The offering two to individuals differing in one or more heritable characteristics.

**Hyperparasite**: A parasitic on another parasite.

**Hyperplasia**: A plant overgrowth due to increased cell division.

**Hypertrophy**: A plant overgrowth due to abnormal cell enlargement.

**Hypha**: A single branch of a mycelium.

**Immune:** Can not be infected by a given pathogen.

**Immunity:** The state o being immune.

**Imperfect fungus**: A fungus that is not known to produce sexual spores.

**Imperfect stage**: The part of the life cycle of a fungus in which no sexual spores are produced. The anamorph stage.

**Incubation period:** The period of time between penetration to a host by a pathogen and the first appearance of symptoms in the host.

**Infection:** The establishment of a parasite within the host plant.

**Infectious disease**: A disease that is caused by a pathogen which can spread from a diseased to a healthy plant.

**Injury**: Damage of a plant by an animal, physical factorsor chemical agent.

**Inoculate:** To bring a pathogen in to contact with a host plant or plant organ.

**Inoculation**: The arrival or transfer of a pathogen on to a host.

**Inoculum**: The pathogen or its parts that can cause infection. The portion of individual pathogens that are brought in to contact with the host.

**Integrated control:** An approach that attempts to use all available methods of control the disease or of all the diseases and pests of a crop plant for best control results but with the least cost and the least damage to the environments.

**Integrated Pest Management:** The attempts to prevent pathogens, insects and weeds from causing economic crop losses by using a variety of management methods that are cost effective and cause least damage to the environment.

***In vitro*:** In culture, outside the host.

***In vivo*:** In the host

**Isolate:** A single spore or culture and the subcultures derived from it. Also used to indicate collections of a pathogen made at different times.

**Isolation**: The separation of a pathogen from its host and its culture on a nutrient medium.

**Life cycle:** The stage or successive stages in the growth and development of an organism that occurs between the appearance and reappearance of the same stage of the organism.

**Monocyclic:** Having one cycle per season.

**Mycelium:** The hypha or mass of hyphae that make up the body of fungus.

**Mycoplasmas:** Pleomorphic prokaryotic microorganism that lacks a cell wall.

**Mycorrhiza**: A symbiotic association of a fungus with the root of higher plants

**Noninfectious disease:** A disease that is caused by an abiotic agent, that is by an environmental factor, not by a pathogen.

**Obligate parasite:** A parasite that in nature can grow and multiply only on or in living organisms.

**Parasexualism:** A mechanism whereby recombination of hereditary properties occurs within fungal heterokaryons.

**Parasite:** An organism living on or in another living organism and obtaining its food from the latter.

**Pathogenecity**: The capability of a pathogen to cause disease.

**Penetration:** The initial invasion of a host by a pathogen.

**Perfect stage**: The sexual stage in the life cycle of a fungus. The telomorph.

**Polycyclic:** Completes many disease/life cycles in one year

**Polyetic:** Requires many years to complete one life or disease cycle.

**Potyvirus:** Potyviruses infect plants and belongs to the family Potyviridae. The virious is non-evveloped, the nuclo capsid is filamortous. Members of this grips may cons. Significant losses in agricultural and horticultural crops

**Primary infection**: The first infection of a plant by the overwintering or oversummering pathogen.

**Primary inoculum**: The overwintering or oversummering of pathogen or its spores that cause primary infection.

**Resistance**: The ability of an organism to exclude or overcome, completely or to some degree, the effect of a pathogen or other damaging factor.

**Resistant:** Possessing qualities that hinder/restrict the development of a given pathogen the material may beInfected little or even not at all.

**Rust:** A disease caused by one of the Uredinales (rust fungi) giving a rusty appearance to a plant.

**Secondary infection**: Any infection caused by inoculum produced as a result of a primary or a subsequent infection; an infection caused by secondary inoculum.

**Secondary inoculum**: Inoculum produced by infections that take place for the first time during the same growing season.

**Sign:** The pathogen or its parts or products seen on a host plant.

**Soil Solarization**: Attempt to reduce or eliminate pathogen populations in the soil by covering the soil with clear plastic so that sun rays can will raise the soil temperature to the levels that kill the pathogens.

**Spiroplasmas:** Pleomorphic, walless microorganisms that are present in the phloem of diseased plants. They are often helical in shape and are thought to be a kind of mycoplasma.

**Spore:** The reproductive unit of fungi consisting of one or more cells, it is analogous to the seed of green plants.

**Sterile fungi**: A group of fungi that are not known to produce any kind of spores.

**Suppressive Soil:** Soils in which certain diseases are suppressed because of the presence of antagonistic microorganism to the pathogen.

**Susceptible:** Lacking the inherent ability to resist disease or attack by a given pathogen; non-immune.

**Susceptibility:** The inability of a plant to resist the effect of a pathogen or other damaging factor.

**Symptoms**: The external and internal reactions or alterations of a plant as a result of a disease.

**Telomorph:** The sexual or so-called perfect growth stage or phase in fungi.

**Teliospore:** The sexual, thick walled resting spore of the rust and smut fungi.

**Tolerance:** The ability of a plant to sustain the effects of a disease without dying or suffering serious injury or crop loss. Also, the amount of toxic residue allowable in or edible plant parts under the law.

**Tospovirus (Bunyaviridae):** Tospoviruses (Family-*Bunyaviridae*) are enveloped isomatic RNA viruses with a tripartite genome containing small (s). medium (M) and large (L) segments of SSRNA. They ae transmitted through Thrips (*Thrips tabaci*) inpropagated manner and are one of the most important plat virus groups infections a wide range of economically important plants worldwide.

**Variability:**The property or ability of an organism to change its characteristics from one generation to the other.

**Vector:** An animal (like, Aphid, white fly, nematode, jassid etc) able to transmit a pathogen.

**Vertical Resistance:** Complete resistance to some races of a pathogen but not to others.

**Virion:** A virus particle.

**Viroid:** Small, low-molecular weights RNA that can infect plant cells, replicate themselves, and cause disease. Virus without coat (naked DMA or RNA in virus)

**Virulence:** The degree of patogenicity of a given pathogen (ability of pathogen to cause disease).

**Virulent:** Capable of causing a severe disease, strongly pathogenic.

**Virus:** A submicroscopic obligate parasite consisting of nucleic acid and protein.

**Virusoid:** The extra-small circular RNA component of some isometric RNA virus.

**Zoospore:** A spore bearing flagella and capable of moving in water.

**Zygospore:** The sexual or resting spore of zygomycetes produced by the fusion of two morphologically similar gametangia.

# Acronyms and Abbreviations

| | | |
|---|---|---|
| AMF | : | Arbuscular Mycrrlizal Fungi |
| a.i. | : | Active ingradient |
| ACGR | : | Annual Compound Growth Rate |
| Ag GDP | : | Agricultural Gross Domestic Product |
| AICRP | : | All India Coordinated Research Project |
| AICRPDA | : | All India Coordinated Research Project for Dry Land Agriculture |
| ASRB | : | Agricultural Scientist Recruitment Board |
| AVRDC | : | Asian Vegetable Research and Development Centre |
| BCA | : | Biocontrol Agent |
| C.O. | : | Causal Organisms |
| CPRI | : | Central Potato Research Institute |
| CRIDA | : | Central Research Institute for Dryland Agriculture |
| CSSRI | : | Central Soil Salinity Research Institute |
| CMI | : | Commonwealth Mycological Institute |
| CSWRI | : | Central Sheep and Wool Research Institute |
| CV | : | Coefficient of Variation |
| DES | : | Directorate of Economics and Statistics |
| DOGR | : | Directorate of Onion and Garlic Research |
| EC | : | Emulsifiable Concentration |
| FAO | : | Food and Agriculture Organization of the United Nations |
| FSCs | : | Farmers Science Centers |
| FTCs | : | Farmers Training Centers |
| FYM | : | Farmyard Manure |
| GA3 | : | Gibberelic Acid |
| GDP | : | Gross Domestic Product |
| HYV | : | High-yielding Varieties |
| IARCs | : | International Agricultural Research Centre's |
| IARI | : | Indian Agricultural Research Institute |
| ICAR | : | Indian Council of Agricultural Research |
| ICRISAT | : | International Crops Research Institute for the Semi-Arid Tropics |
| ICMR | : | Indian Council of Medical Research |
| IDM | : | Integrated Disease Management |
| IPI | : | Input Price Index |
| IPCC | : | Intergovernmental Panel on Climate Change |
| IPM | : | Integrated Pest Management |
| IRR | : | Internal Rate of Return |
| ITCC | : | Indian Type Collection Centre |

| | | |
|---|---|---|
| IIVR | : | Indian Institute of Vegetable Research, Varanasi |
| IYSV | : | Iris Yellow Spot Virus |
| NARS | : | National Agricultural Research System |
| NATP | : | National Agricultural Technology Project |
| NCIPM | : | National Center for Integrated Pest Management |
| NHRDF | : | National Horticulture Research and Development Foundation |
| NIPHM | : | National Institute of Plant Health Management |
| NGOs | : | Non-Governmental Organizations |
| NSC | : | National Seeds Corporation |
| NSKE | : | Neem Seed Kernel Extract |
| OPVs | : | Open Pollinated Varieties |
| OYDV | : | Onion Yellow Dwarf Virus |
| PHT | : | Post Harvest Technology |
| PTM | : | Potato Tuber Moth |
| R&D | : | Research and Development |
| R&E | : | Research and Extension |
| RKN | : | Root Knot Nematode |
| RCP | : | Real Cost of Production |
| SAUs | : | State Agricultural Universities |
| TERI | : | The Energy and Resources Institute |
| TH | : | Trichoderma harzianum |
| TV | : | Trichoderma viride |
| TPS | : | True Potato Seed |
| UNDP | : | United Nations Development Programme |
| VC | : | Vermicompost |
| WP | : | Wettable Powder |
| WTO | : | World Trade Organization |
| WHO | : | World Health Organization |

# Introduction

Vegetables are important source of dietary fibers, minerals, antioxidants and vitamins. Shifting from a non-vegetarian diet to vegetarian, global recognition of the importance of vegetables for human health and their medicinal and nutritional value have contributed to a steady upward trend in vegetable production. China is ranked first in the world and currently produces around 237 million tons of vegetable. The total vegetable production of India during the year 2009-10 was approximately 90 million tons. India has made a quantum leap in vegetable production, securing second position in the world. Other major vegetable producing countries are Japan, the Republic of south Korea and Iran.

Pesticides are valuable tools in vegetable production, but unfortunately they are often being used irresponsibly, causing hazards to producers, consumers and the environment. In addition, they can exacerbate pest and disease problems by inducing resistance and suppressing the natural enemies which exert natural biological control.

This book aims at promoting an integration of appropriate chemical, cultural, physical, genetic and biological technologies in to an Integrated Pest Management

(IPM) strategy for vegetable crop protection. Adoption of this approach will empower farmers with effective pest and disease management methods and reduce reliance on synthetic pesticides gradually, resulting in more sustainable vegetable production with higher yields of safer produce.

For general understanding IPM can be defined as a long-standing, science-based, decision-making process that identifies and reduces risks from pests and pest management related strategies. It coordinates the use of pest biology, environmental information, and available technology to prevent unacceptable levels of pest damage by the most economical means, while posing the least possible risk to people, property, resources, and the environment. IPM provides an effective strategy for managing pests in all arenas from developed residential and public areas to wild lands. IPM serves as an umbrella to provide an effective, all encompassing, low-risk approach to protect resources and people from pests.

Pests may be an insect, nematode, weed, plant pathogen or vertebrate that reduces crop yields, negatively impacts animal or human health or causes structural damage.

## Origin of the term, IPM

The phrase, Integrated Pest Management, was found used in entomology literature in 1967. It was based on a concept termed integrated control, which had been previously developed by entomologists. The exact term integrated pest management and the IPM abbreviation entered the public arena and found its way in to mainstream scientific literature in 1972 in both, a paper from President Nixon's administration to congress and in a report entitled Integrated Pest Management of the Council on Environmental Quality.

IPM is often mistakenly associated only with entomology and that too only for insect pests. However, in contrast to entomology, plant pathology has, since its beginning, approached plant pests through multiple, or integrated strategies. The terminology traditionally used by plant pathologists was typically tactics. These traditionally-used tactics included rotation, pathogen-free seed, and genetic host resistance. However, in 1975 the federal USDA Cooperative Extension Service (CES) funded IPM programmes in every state and at this point the term, IPM became commonly used in plant pathology literature also.

Kogan in as general article, *IPM History*, detailed the components of this abbreviation, IPM, as it was generally accepted during the time of its entry in to scientific and public nomenclature in the early 1970s as:

1. *Integration* meant the harmonious use of multiple methods to control single pests as well as the impacts of multiple pests.

2. *Pests* were any organism which was detrimental to humans, including invertebrate and vertebrate animals, pathogens and weeds.
3. *Management* referred to a set of decision rules based on ecological principles, economic and social considerations.

Kogan also noted that IPM was to address a multidisciplinary approach (e.g. Entomology, Plant pathology, Nematology, Weed science, etc.).

## Development of the IPM Concept

Before World War II and the accompanying widespread use of organo-synthetic chemicals, agricultural pest problems were commonly managed using cultural practices and information on pest life cycles and biology. However, development and use of organo-synthetic chemicals changed the focus to pest management by chemical means.

During the late 1950s and early 1960s increasing levels of chemical resistance in insect populations and pesticide injury to non-target organisms became evident and caused concern. These problems were brought about by reliance on insecticides for insect control. Entomologists began to develop and foster the concept of IPM. Ecological issues relating to the use of pesticides were also emerging during this time and also highlighted the need for a re-evaluation of reliance on chemical pest control. Here comes the issues and impressions raised by Rachel Carlson in her famous book entitled "Silent Spring". Silent Spring can be considered as Harbinger of integrated pest management in the ecosystem. Meanwhile new developments in computer science, along with the emerging sciences of operations research, systems analysis, and modeling, allowed more sophisticated, multi-level, integrated approaches to pest management to emerge. Therefore, the concept of IPM began to gain interest among researchers, extension personnel, agricultural stakeholders, the federal government and the public at large.

## Principles of IPM

- Survey and surveillance
- Grow a healthy crop
- Preserve natural enemies

- Visit and observe fields regularly
- Farmers become the experts

## Integrated Pest and Disease Management: Key Strategies

The success of integrated diseases and pests management approach in vegetable crops is only possible when we apply all the following practices and precautions every year on community basis in large area.

- Avoid indiscriminate use of fungicides and insecticides and do not mix incompatible pesticides, phytohormones and micronutrients at a time.
- Soil as a reservoir of harmful as well beneficial microorganisms, therefore soil health should be properly managed by timely tillage, summer ploughing, green manuring, optimum C: N ratios, balanced macro and micronutrients, aeration etc.
- Summer ploughing followed by irrigation in summer and then subsequent ploughing is highly effective.
- Seed health must be maintained up to prescribed standard and ensure that seed should be free from internal and external pathogen, contamination of any infected crop debris, sclerotial bodies etc.
- Field sanitation as essential practices where removal and burning of infected crop debris, alternate and collateral weed hosts should be carried out periodically.
- Crop rotation is very important for soil borne diseases where non-host crop preferably cereals should be selected for particular pathogen.
- Avoid off season vegetables and intensive cropping because it prolongs the perpetuation period of a pathogen and do not break the life cycle of the pathogen.
- Knowledge of correct diagnosis of the diseases and disease cycle of the pathogen is very important to apply different methods of critical phase of life cycle for maximum efficacy.
- Application of biological agents particularly resident antagonist accompanied with green manuring, FYM or any organic matter should be maximized.
- Always prefer to grow tolerant varieties. However, it is difficult to get resistance against diseases with all desirable traits in hybrid

- Proper post harvest management practices should be followed to avoid any rotting of vegetables during storage and transportation.

Knowledge of the following aspects of disease development is essential for effective and economic management:

- Cause of the diseases.
- Mode of perennation and dissemination of the infectious causes.
- Host-parasite relationship and means of secondary spread.
- Effect of environment on pathogenesis in the plant and spread of the disease in the plant population.

# CONTENTS

# CHAPTER – 1

# Diseases of Allium Crops — Onion and Garlic

## 1. Damping-off

### Pathogen

*Pythium aphanidermatum* (Edson.) Fitz.
*P. debaryanum*
*Rhizoctonia solani* Kuhn.
*Fusarium spp.*
*Sclerotium spp.*

### Symptoms

Damping-off of onion seedlings occurs in two stages:

(i) **Pre emergence**
(ii) **Post emergence**

In pre emergence stage, plumules are killed before they reach the soil surface. They may be infected, killed even before the hypocotyls have broken the seed coat. The radicle and plumule, when come out of the seed, undergo complete rotting. *Since this happen below the soil surface and the disease often not*

**Figure 1:** Onion seedlings infested with damping off

*noticed, the failure in emergence of seedlings is attributed to the poor quality of seeds.* The post emergence damping off is characterized by the toppling over of infected seedlings anytime after they emerge from the soil. It usually occurs at the ground level and infected tissues become soft and water soaked. As the disease advances, the shoot becomes constricted at the base and plant collapses. *It is observed that most of the loss occurs due to pre emergence damping-off.* It is more common during *kharif* season when temperature and humidity remain very high.

## Management

- Proper drainage of field is most essential.
- Sowing of clean & uninfected seed on raised beds.
- Crop rotation with non host like cereals.
- Soil solarization of nursery beds with transparent polythene (LLDPE) for 30 days before sowing gives good management of damping-off.
- Seed treatment with Thiram, Captan @ 2.5 g/kg seed or Carbendazim @ 0.1% or *Trichoderma viride* @ 4-5 g/kg are effective.
- Drenching the nursery bed with Thiram @ 2.5% or Carbendazim @ 0.1% or *Trichoderma viride* @ 4-5 g/litre.

## 2. Purple blotch

### Pathogen

*Alternaria porri* (Ell.) Neerg.

### Symptoms

The disease appears on the leaves as small, whitish, sunken lesions. These lesions later on enlarge and eventually encircle the whole leaf. Later on, oval shaped dark zones appear on the surface of the leaves, retaining the **characteristic purple colour**. The leaves and shoots fall down or lodge gradually. Concentric zones may develop within the lesions. Disease is favoured

**Figure:** Crops affected with Purple Blotch **(a)** Onion **(b)** Garlic **(c)** Onion Seed Crops Infected with Purple blotch

by heavy dew in desert areas and by foggy and rainy weather in other regions. The spores are air borne in nature.

### Management

- Use healthy seeds.
- Crop rotation for 2-3 years with non-host crops should be followed.
- Spraying of Mancozeb @ 2.5g/litre water and Chlorothalonil @ 2.0g/litre water should be sprayed at 15 days intervals.
- Spraying of *Trichoderma viride* suspension @ 5g/litre at weekly intervals would be effective against this disease.
- Seed treatment with Captan @ 2.5 g/kg seed, Thiram @ 2.5 g/kg or *Trichoderma viride* @ 4-5 g/kg seed before nursery sowing.
- Summer ploughing has been found to reduce the incidence of Purple blotch.
- Growing of moderately resistant cultivars like, G-1, G-50, and G-323 are recommended

## 3. Stemphylium blight

### Pathogen

*Stemphylium vesicarium* Wallr. (Sacc.)

### Symptoms

The disease symptoms appear as small, yellow to orange flecks or streaks on the leaf. These streaks flacks soon develop in to elongated, spindle shaped to ovate or elongate, diffused spots, often reaching the leaf tips. They usually turn grey at the centre, brown to dark olive brow in margins. These spots frequently coalesce in to extended patches, causing blight of the leaves and gradually covering the entire foliage. In advanced stages, these lesions may engirdle and kill leaves and seed stalks.

### Management

- Spraying of Mancozeb @ 2.5g/litre water along with sticker is effective to manage the disease. Spraying has to be done just after appearance of disease and same may be repeated fortnightly.
- Spraying of Sixer (Metalaxyl + Mancozeb) @ 2.5g/litre of water with sticker at 15 days intervals also does wonder.
- Crop rotation with other vegetable non-host crops like, tomato, capsicum, pea, and cereals like wheat and paddy reduce the disease severity.

**Figure** Stemphylium blight affected Onion **(a)** and Garlic **(b)** crops

- Deep summer ploughing also reduces the disease incidence.

## 4. Cercospora leaf spot

### Pathogen

*Cercospora duddiae* Walles.

### Symptoms

The disease appears on the leaves as small ash coloured and irregular shaped spots on leaf lamina. The spots coalesce gradually, resulting in blighting of the

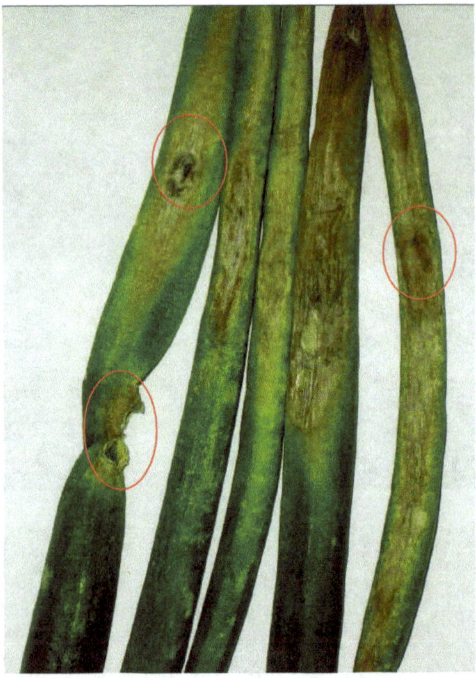

**Figure:** Cercospora leaf spot infected onion leaves

foliage. High temperature prolonged wet conditions favour the disease development.

### Management

- Spraying of Mancozeb @ 2.5g/litre water and Copper Oxychloride @ 3.0g/litre water at 15 days intervals give good results.
- Spraying of Chlorothalonil (Kavach) @ 2.0 g/litre water and Sixer (Metalaxyl + Mancozeb) @ 2.5 g/litre water at 15 days intervals give good management.
- Deep summer ploughing.

## 5. Powdery Mildew

### Pathogen

*Leveillula taurica* (Lev.) Arnaud

### Symptoms

Whitish, circular to oblong lesions of variable sizes appear on abaxial surface of

**Fig** Whitish powdery mass on onion leaves/plants (Source: www.usda.org)

leaves and older leaves are infected first. The lesions are covered with white powdery mass of fine hyphae and conidia.

### Management

- Use of sulphur fungicides eg. Sulfex, wettable Sulpher, Carathene etc. @ 0.2% at fortnightly intervals.
- Proper drainage is essential for prevention of this disease.

## 6. Downy Mildew

### Pathogen

*Pernospora destructor*

### Symptoms

The pathogen can infect both onion and garlic. The first indication of appearance of this disease is a fine, furry, greyish white to purple growth on the surface of older leaves. Leaf tissue, under the growth, becomes pale green, then yellow, and finally collapses. Large, yellowish and circular clumps of infected plants measuring a few too many feet in diameter. The yellowing patterns often enlarge in the direction of prevailing winds. Disease can develop from an initial infection by air borne spores into an epidemic formvery quickly, if humidity and

**Fig.** Downy mildew affected onion leaves

temperature conditions are favuorable. Initial inoculum of disease can infect bulbs, sets, seeds and plant debris.

## Management

- Use disease free seeds, bulbs and sets.
- Follow 3-4 years crop rotation with other non-host crops like vegetables and cereals.
- Grow eed onion cultivars that are moderately resistant to downy mildew.

- Spraying of Mancozeb @ 2.5 g/litre at 15 days intervals
- Always avoide overwatering in the field.

## 7. Rust

**Pathogen** *Puccinia porri*

### Symptoms

Rust is primarily a disease of garlic, although onion, leeks, shallots and wild species of *Allium* are hosts. Small, reddish to dull orange oval shaped pustules develop on leaf blades. Reddish air borne urediospores are copiously produced within the lesions. Later in the growing season, the lesions may appear dark because black teliospores develop within the pustules. Heavily infected leaves turn yellow and may collapse prematurely. Rust decreases the leaf area surface

**Figure:** Rust affected garlic plant

for photosynthesis. When infection is severe, bulb size and quality are reduced. It is a sporadic disease that generally causes little or no economic damage.

### Management

- Plough under and busy in soil infected crop residues.
- Selecting planting dates and scheduling of irrigation to avoid long periods of leaf wetness when temperatures are highly beneficial.

- Avoid over application of nitrogenous fertilizers and ensure adequate potassium fertilization.
- Spraying of Propiconazole and Hexaconazole @ 0.1% at fortnightly intervals gives good management.

## 8. Botrytis Leaf spot

### Pathogen

*Botrytis ceneria*

### Symptoms

Botrytis leaf spot occurs on onion but till date has not yet been reported on garlic. The leaf spot lesion initially has a small white, necrotic centre surrounded by a light green halo. The fungus survives in old bulbs or onion leaf debris as sclerotia and mycelia where as in soil as sclerotia. Spores produced in infested soil or plant debris are carried to onion leaves by wind or splashing water.

### Management

- Cultural practices that promote free air movement and rapid drying of leaves can help reduce infections and disease severity.
- Crop rotation with non-susceptible crops such as alfalfa or small grains cereasl/milletssuppresses the build-up of sclerotia in soil.

## 9. Anthracnose disease (Twister)

### Pathogen

*Colletotrichum gloeosporioides* Penz.

### Perfect Stage as Sexual stage

*Glomerella cingulata* (Stonem) Schrenk & Spauld

### Symptoms

Curling and twisting of leaves, chlorosis and abnormal elongation of neck portion of the shoots are some of the major symptoms. Bulb development is also affected. White oval sunken lesions are formed on the leaves. In advance stage, roots become sparse and plant may die. Bulbs are slender and may rot before harvest or during initial period of storage.

**Figure:** Curling and twisting of onion plant

### Management

- Use of good quality seeds/bulbs/cloves.
- Seed treatment with Thiram or Capton @ 2.0-2.5 g/kg seed before nursery sowing.
- Seed treatment with *Trichoderma viride* @ 5.0 g/kg seed.
- Soil treatment with Benomyl or Benlate (@ 2.0 g/m$^2$ and spraying of Mancozeb @ 2.5 g/litre water are effective.

- Sanitation and destruction of infected plants debris help in reducing the disease incidence and inculum load for the next crop.

## 9. Southern blight

### Pathogen

*Sclerotium rolfsii* Sacc.

### Symptoms

The early symptom chlorotic dieback is similar to those of caused by *Fusarium*. There is characteristic growth of pale mycelium on the soil surface around the shoot base on which sclerotia are formed. It is active only on the upper surface of soil and is dependent for its activity on the presence of dead or senescent plant tissue.

### Management

- Crop rotation with cereals to reduce the sclerotial population.
- Soil solarization reduces sclerotial population.
- Soil treatment with *Trichoderma viride* also reduces the infection.

## 10. White rot

### Pathogen

*Selerotium cepivorum* Berk.

### Symptoms

Pathogen is soil inhabiting and invades the roots and the basal part of the bulb scales. The first symptom of the disease is premature yellowing and dying of older leaves, stunting of the plants and leaf tip burn, followed by destruction of the root system and shoot dieback as well as rotting of the bulb. The roots are generally destroyed and there are semi watery decay of the scale with abundance of superficial white mycelium. Brown to black sclerotia are developed on surface or within the tissues. Sclerotia can survive for more than 20 years, even in the absence of a host plant. Disease severity depends on sclerotia levels in the soil at planting stage. Disease development is favoured by cool and moist soil conditions. The soil temperature range for infection is 50° to 75° F, with optimum 60° to 65° F.

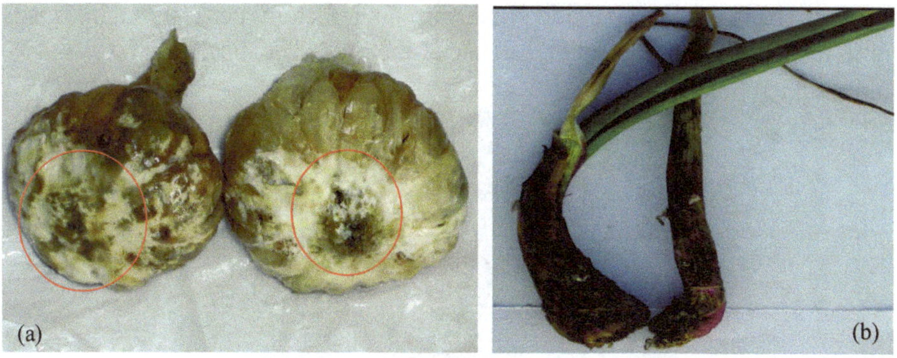

**Figure:** White rot affected **(a)** Garlic bulbs and **(b)** Onion plants

## Management

- Removal of infected plants during on season reduces the sclerotial population and also avoids incorporation of the same to the soil.
- Solarization of soil at high temperature, i.e. 45°C for 18 hours or 45°C for 6 hours reduces the incidence.
- Spraying of Iprodione (Rovral) @ 0.25% was effective and Benomyl @ 0.1% have reduced disease incidence.
- *Trichoderma viride*; *Gliocladium zeae*, *Penicillium nigricans, Bacillus subtilus* and *Trichoderma harzianum* are highly effective against post-harvest diseases like this.
- Use of clean seeds, cloves and bulbs for planting/transplanting.
- Light irrigations also help in minimizing the white rot incidence.

## 11. Basal Rot/Bottom Rot

### Pathogen

*Fusarium oxysporum* f. sp. *cepae* Schlecht.

### Symptoms

The main symptoms of disease in onion are delayed emergence of seedlings, seedling damping off, stunted growth and basal rotting in growing plant accompanied by early maturation and basal rotting in store. Roots may turn pink and gradually decay until the entire root disappearer. In garlic, pre emergence decay of cloves, stem plate and post harvest decay of cloves in stored bulbs are main symptoms. Infected garlic show reddish or reddish purple discolouration on shoot and bulb early in the season with some discolouration on bulb sheath at harvest. The fungus survives indefinitely in soil. Infections occur through wounds or in the vicinity of old root at the base of the bulb. The disease is favoured by high soil temperature.

**Fig:** Basal Rot affected onion crop **(a)** and **(b)** bulb **(c)** Garlic

### Management

- Proper drainage is essential during entire cropping period.
- Proper drying and curing are essential before storage.
- Use of bioagents like *Trichoderma viride*, *Pseudomonas* spp., *Beauveria bassiana* with vermicompost before transplanting is effective against this disease.
- Deep ploughing and avoiding injury during cultural practices are beneficial.
- Avoid growing onion in the fields which are knowm for having a history of basal rot problem in past years and rotate 3 to 4 years with other non host crops.
- Storage of onion and garlic at cool temperatures should be resorted to since infection is favoured by warm condition.
- Removal of infected plants and planting of disease free seed, are best proposition.

## 12. Brown rot

### Pathogen

*Pseudomonas aeruginosa*

### Symptoms

Dark brown discolouration in bulb scale is the characteristic symptom of this disease. Browning of inner scale along with rotting is the main symptom of this disease. The rotting starts from the inner scale and spread to outer scales. Apparently, the bulb seems to be healthy, but when pressed, the white oozing cloud of bacterial mass noticed from the neck. In several cases, the whole lot of bulbs get rotten giving the typical odour in storage.

**Figure:** Brown Rot affected onion bulb

### Management

- Proper curing of bulb is required before storage.
- Use of Maleic Hydrazide (a hormone) and Isopropyl Phenyl Carbamate (IPC) @ 20 ml/litre before one month of harvest are most important.
- Neck cutting at about 2.5-3.0 cm long above the bulb is found to reduce the bacterial infection.
- Light irrigation is required during entire cropping period.

## 13. Soft Rot

### Pathogen

*Erwinia caratovora*

### Symptoms

Severe discolouration with soft rotting with water soaking of one or more of the inner fleshy scales of onion bulbs are the main symptoms of this disease. Severe

**Figure:** Soft rot infected onion (a) and garlic bulb (b)

infection occurs at high temperature. The affected fleshy scaled tissues are water soaked and pale yellow to light brown in colour and become soft as the rotting progresses. The whole bulb may break down and a watery liquid (bacterial wars) may ooze out having foul odour as disease advance if squeezed. The organisms generally appear just before or at the time of harvest or in the storage. The pathogens are soil borne in nature and may spread with irrigation water.

### Management

- Proper curing is required after harvest.
- Proper drainage is required during cropping period.
- Reduce the doses of nitrogenous fertilizers.
- Harvest only when onions tops are well matured and about to dry.

## 14. Neck Rot

### Pathogen

*Botrytis allii* Pers.

### Symptoms

Infection usually takes place through neck tissues and occasionally through other means. The first sign is the softening of the affected tissues which takes a sunken hooked shaped appearance. At later stage, decaying starts, but remains separated from healthy tissue by a definite margin. A dense greyish mycelial mat develops upon the decaying tissue of the scope bearing short conidiophores with conidia. The disease progresses rapidly down the scales of neck tissue.

**Figure:** Severe rotton in the neck region of onion bulb

Black sclerotial bodies are developed later. When diseased bulbs are cut open, water soaked brown tissues are seen near the neck region.

### Management

- Seed treatment with Benomyl @1 g/kg seed or Benomyl + Thiram @ 1 g/kg seed reduce the disease incidence.
- Pre harvest sprays of Benomyl @ 0.1% reduce the fungal infection.
- Proper curing is required before storage
- Necks cutting at 2.5-3.0 cm height above the bulb reduces the infection.
- Use disease free planting materials.

## 15. Pink Root Rot

### Pathogen

*Pyrenochaeta terrestris* de Not.

### Symptoms

The characteristic symptom of Pink Root rot is the pink colouration of the roots. The affected roots initially turn yellow but later on become soft and ultimately take distinctly deeper pink colour. The new roots which grow from the infected plants get immediately infected and become functionless. The affected plants are not killed but development is retarded as leaf number and size are reduced. Bulbing starts earlier, but the size is reduced at maturity. The disease is confined to the roots only. The fungus is a common soil inhabitant that penetrates onion and garlic roots directly, wounds are not necessary for infection, but weak plants are more susceptible. It can spread with water and along with dirty equipments.

**Figure:** Pink root affected onion bulb/roots

### Management

- 5-6 years crop rotation gives good control of pink root.
- Soil solarization also helps in reducing the disease.
- Use of Dichloropropene @ 450 litre/ha increases the yield.
- Grow resistant cultivars

## 16. Black Mould

### Pathogen

*Aspergillus niger* van Tiegh.

### Symptoms

Black mould occurs on both the crops, onions and garlic. Fungus infects the neck of bulbs and occasionally penetrates from side and base of the bulb, where damage to the dry skins has exposed the bulb scales. In advance stages of the symptoms, the entire surface of bulbs turns black and all scales are infected. The fungus survives on decaying organic matter such as plant debris. It reduces the market value of the bulb.

**Fig:** Black powdery mass on **(a)** onion and **(b)** garlic bulb

### Management

- Proper drying and curing of the bulbs after harvest are of utmost important.
- Bulbs should be dusted with Calcium Carbonate should be godowns are fumigated with Nitrogen Trichloride @ 430 mg/m$^2$.
- There are no chemicals for instant management of black mould fungus.
- Handling of bulbs to avoid bruising can also reduce injury and invasion of the fungus.

## 17. Blue mould

### Pathogen

*Penicillium corymbiferum*. Link ex Fr.

### Symptoms

Blue mould generally appears during harvesting and storage. The symptoms of the disease are pale yellowish blemishes, watery soft spots, or occasionally a purplish red stain on the scales. A green to blue green mold may develop on the surface of lesions when bulbs are cut longitudinally, one or more of the fleshy scales may appear water soaked and exhibit a light tan or grey colour. In advanced stages infected bulbs may disintegrate in to watery rot.

**Figure:** Garlic bulbs affected with Blue Mould

### Management

- Proper drying and curing of bulbs after harvest and before storage.
- Injury to bulbs should be avoided during post harvest handling.
- Bulbs should be stored at temperature of 5°C or less with low relative humidity.

## 18. Onion Smut

### Pathogen

*Urocystis cepulae* (Frost.) Kolar Dt.

### Symptoms

The disease appears first on the cotyledons of the young plants soon after they emerges from the seed or bulb in soil. The lesion consists of dark slightly thickened area involving leaf or cotyledon from one to several millimetres. Some times, the major part of a leaf has single lesion and in such cases the leaf tends to curve downward abnormally. The lesions on plants break open and expose a black powdery mass of spores. The fungus does not produce rot in storage, but the affected bulb may be subjected to invasion by other secondary pathogens.

### Management

- Seed treatment with Thiram @ 2.5g/kg was found most effective to manage the disease.

**Figure:** Smut affected onion bulb

- Application of Formaldehyde solution 28 ml/litre of water per 30 mt square iarea s effective in managing the disease.
- Use of phosphatic and potassic fertilizers also helps in reducing smut infection.

## 19. Sour Skin

### Pathogen

*Pseudomonas (Burkhcepacia) cepacia*

### Symptoms

Affected outer scales become yellow and slimy giving vinegary smell. Externally, bulbs seem sound but the neck region may softened. In advance stages, healthy, scales can slip off during handling. Young leaves sometimes die back. The optimum temperature for the growth of the fungus ranges between 30 and 35°C. No growth occurs at 40 °C. Inoculum build up of *P. cepacia* has been associated with contaminated water.

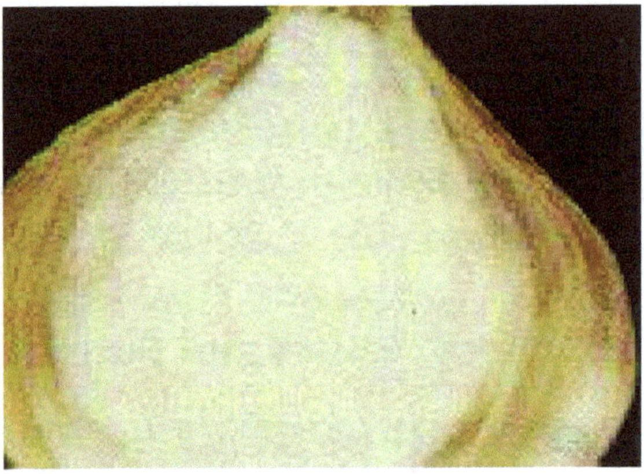

**Figure:** Onion bulb infected with sour skin

### Management

- Use of balanced doses of manures and fertilizers are highly required.
- Proper drainage is required
- Crop rotation with non-host vegetables like, tomato, capsicum, pea and cereals like, wheat and paddy may be employed.
- Proper curing is essential after the harvest and before storage.

## 20. Bulb Canker/Skin blotch

### Pathogen

*Embellissia allii.*

### Symptoms

The initial symptoms of the disease are small blackish dark spots on the outer scales of the bulbs, which later on enlarge and cover whole bulbs with black powdery mass. Under advance conditions, decay of the cloves has been observed and ultimately reduces the market value.

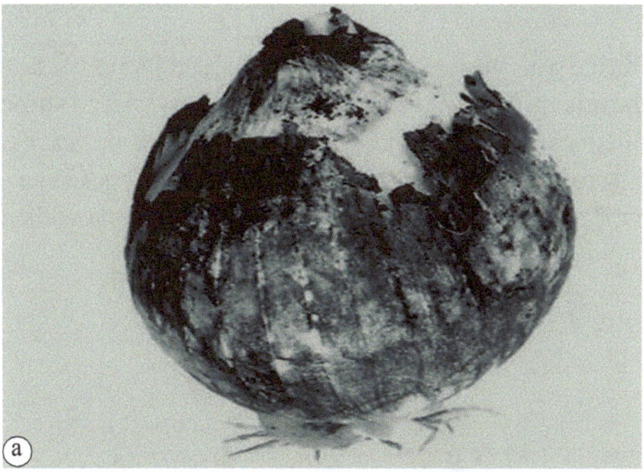

**Figure:** Onion **(a)** and garlic **(b)** bulbs affected with canker/skin blotch

### Management

- Spray of Carbendazim @ 1.0 g/litre should be given before 15-20 days of harvest of the crops.
- Seeds (Cloves) should be treated with bavistine @ 1.0 g/kg seed before planting.
- Proper curing of bulbs should be taken before storage.

## 21. Garlic Mosaic

### Pathogen

Virus

### Symptoms

Garlic plants infected with mosaic virus show typical symptoms of chlorotic mottling and stripes on the emerging leaf followed by pale yellow broken stripes resulting in typical mosaic pattern on matured leaves. Yellowish dots on leaves, whitish leaf margin, or twisting of leaves are also observed on some cultivars. Generally symptoms are more pronounced in young leaves. Infected plants are stunted thereby bulb size is reduced.

**Figure:** Mosaic infected garlic leaves

### Management

- Spraying of Monocrotophas @ 0.05%, Endosulphan @ 0.25% or Methyl Dematon @ 0.075% at 10-15 days intervals
- Treatment of Cloves with Bavistin @ 0.1% before planting is also recommended
- Use of HNPV and some neem based botanicals at weekly intervals has been found beneficial
- Virus free stocks should be produced from meristem tip culture and multiplied in the areas which are vectors of commercial garlic cultivation is of utmost importance to prevent reinfection by insects (vectors).

## 22. Iris Yellow Spot Virus (IYSV)

### Pathogen

Virus (*Tospovirus*)

Vector: *Thrips tabaci* Lindeman

### Symptoms

The disease symptoms vary among onion bulbs crop and seed crops, but often appear as straw-colored, diamond shaped lesions on leaves and scapes with twisting or banding flower bearing stalks are observed in onion plants. Some lesions have distinct green centre with yellow or tan borders, other lesions appear as concentric rings of alternating green and yellow/tan tissue. Infected plants are found scattered or generalized throughout a field. Large necrotic region may develop on scapes and cause collapse of the scape. Diseased plants may be scattered or widespread across a field, but the highest incidence of disease is often found on the edges of the field. The virus infects most of the *Allium* species worldwide. It also infects some ornamental plants like, *Iris, Lisianthus, Calendula* and some weeds like, *Jimson wed, chenopodium, tobacco redroot* and *pigweed.*

### Management

- Iris Yellow Spot Virus (IYSV) is a Topso virus, similar to that of Tomato Spotted Wilt Virus, which is currently thought to be vectored solely by onion thrips (*Thrips tabaci*), so alternate spray of softer insecticides formulations (Spinosad, Neem extract) and organic mulches (straw) are effective against IYSV.

**Figure:** Onion bulb **(a)** seed crop **(b)** and **(c)** Garlic plant infected with IYSV

- Spraying of Deltamethrin and Fipronil @ 1.0 ml/L of water after best mangement of this disease.
- An integrated approach is essential for management of IYSV.
- Eliminate weeds in and around onion fields, especially volunteer onions and wild alliums.

## 23. Onion Yellow Dwarf Virus (OYDV)

### Pathogen

Virus (Potyvirus)

### Symptoms

Onion Yellow Dwarf Virus (OYDV), an aphid-borne potyvirus, is one of the major viral pathogens of onion and garlic. It restricts seed production in both the crops. Leaves of OYDV infected onion show irregular yellow striping to almost complete yellowing and also downward curling, flattening and crinkling following by stunting. In garlic, OYDV produces symptoms depending on virus isolate and type of cultivars. Reduction in growth and bulb size also occurs. Infection by other viruses such as leek yellow stripe virus, garlic common latent virus and shallot latent virus also occur and may aggravate the symptoms. However, OYDV is recognized as a major element of the virus disease complex in garlic. It spread from plant to plant by the green peach aphid, *Myzus persicae*, and other aphids in a non persistent manner.

**Figure:** OYDV Infested garlic plants

### Management

- Removal of virus infected plants.
- Alternate spray of systemic insecticides with neem based botanicals.
- Collection of healthy seed from disease free plants.
- Use of virus free planting stock.

## 24. Rot-knot nematode

### Pathogen

*M. javanica, M. chitwoodi*

### Symptoms

The root-knot nematode is widely distributed, and various species are obligate parasites with wide host range. Typical symptoms include round spindle shaped swelling on roots. The galls are smaller on onion (1-2 mm in diameter). The infected secondary root systems are shorter and have less roots and root hairs. Plant stunting, poor plant stand and yellowing are also associated with the infestation of root-knot nematode. The optimum temperature of 15-30 °C are suitable for infestation and life cycle of the nematode. It is more severe in sandy textured soil.

### Management Strategies

- Use certified nematode-free seeds.
- Follow three-four years' crop rotation with other non-host vegetables and cereals.
- Eliminate weeds to control the population of nematodes.
- Deep summer ploughing can reduce the nematode population in soil.
- Fumigate onion seeds with Methyl Bromide at the time of nursery raising.
- Use Phorate @ 10 kg/ha or Neemagon @ 10-15 kg/ha at the time of transplantation of onion and garlic.

## 25. Rice Root-knot nematode

### Pathogen

*Meloidogyne graminicola*

### Symptoms

Several plants of red onion showed yellowing of leaves, and these plants wilted during the day but recovered their turgidity at night. Diseased plants are easily pulled from the soil due to the almost complete destruction of the root systems. Numerous galls (ageraging 15-20 galls/plant) are found on affected roots, compared with healthy plants. The size of each gall is around 1.0-1.5 cm, and they are round to oval in shape.

The morphological characters of the second-stage juveniles of the root-knot nematode are examined under light microscope. The average size of eggs is 90.4 x 38.2 mm. The size of male and female were 440 x 16.6 mm. The sizes of male and female are 1221 × 32 mm and 535 × 380 mm, respectively.

**Figure:** Rice root knot nematode of onion

**Figure:** Rice root knot nematode in garlic

### Management Strategies

- Crop rotation with other solanaceous and cucurbitaceous vegetable can reduce the nematode population in onion and garlic.
- Apply vermicompost @ 2-3 tonnes/ha and neem cake at the time transplantation
- Avoid excess irrigation.

## 26. Stem and Bulb Nematode

### Pathogen

*Ditylenchus dipsaci*

### Symptoms

The invaded cotyledon may swell and seedling may collapse. After 10 days of invasion, leaves are deformed and stem splitting takes place. When the nematode multiplies in the leaves, the bulb becomes swollen and distorted, giving bloat like appearance. Auxilliary buds may sprout creating multistemmed plants, which may be swollen and stubby. The inner leaves may coil in the leaf sheath and fail to emerge completely.

### Management

- Crop rotation with other non host vegetables and cereals.
- Elimination of weeds to contain the population of nematodes.
- Fumigation of onion seed and bulbs with Methyl bromicle bromide.
- Deep summer ploughing.
- Use of Phorate @ 10 kg/ha or Nemagon @ 10-15 kg/ha at the time of transplanting.

**Figure:** Garlic bulb infected with stem nematode

# CHAPTER – 2

# Diseases of Leguminous Crops

## Vegetables

## Peas

### 1. Rust

#### Pathogen

*Uromyces fabae* (Pers.) de Bary

#### Symptoms

The fungus produces characteristics rust pustules on the host. Pustules are mostly found on the leaf blade, though often they may be found on the petiole and stem of pea plants. The sori are small, roundish, open, and powdery as well as brown coloured. In severe infections, the leaf may wither, resulting in considerable damage to the crop. As the plants near maturity, the fungus produces the telial stage on the host.

**Figure:** Pea plants **(a)** and leaves **(b)** infected with *Uromyces fabae*

### Management

- Avoid delay in sowing and must complete sowing by first fortnight of October.
- Spraying of Sulfex @ 2.0 g/litre and Propiconazole (Tilt) @ 0.1% at 10 to 15 days intervals.
- Grow disease resistant/tolerant cultivars
- Rotate pea with non-host crops.
- Avoid over application of nitrogenous fertilizers to reduce the rust infection

## 2. Powdery mildew

### Pathogen

*Erysiphe pisi* DC.

### Symptoms

This disease is widespread in nature and economically important all over the world. Symptoms first appear in the month of February as white to light grey, powdery fungal spores on leaves, twigs, tendrils and pods as pustules. Infection of pods lead partial filling of the grain and pods remain smaller in size. The absence of rain and presence of at least slight dew favour disease development.

**Figure:** Powdery mildew infected pea leaves

### Management

- Infected crop debris must be burnt after harvesting of pods.
- Foliar spray of Sulfex @ 0.2%, Tridemorph @ 0.1% or Dinocap @ 0.1% at 5-7 days intervals
- Always use resistant/tolerant cultivars like, NDVP-250, NDVP-4, KS-245, JP-3, JP-4 and Pant P-9
- Avoid excess irrigation in the field

## 3. White rot or Sclerotinia rot

### Pathogen

*Sclerotinia sclerotiorum* de Bary.

### Symptoms

Symptoms are observed as wet, soft and white rotting of the tissues. Very soon white fungus growth is observed on the rotted portion. Later on embedded sclerotia in white mycelium formed on the infected portion as well as inner portion of pith and fruits. Primary infection on pea always starts from floral infection followed by pods.

### Management

- Seed cleaning is essential to remove the sclerotial mixer during threshing and processing period.
- Deep ploughing, low lands with continuous stagnation of water reduce the sclerotial population.

**Figure:** White Rot infected pea plants

- Alternate spray of recommended fungicides at 10-15 days intervals is essential.
- All infected plant parts must be collected and burnt before drying of the plants.
- Removal of the weeds from the fields.

## 4. Aschochyta blight

### Pathogen

*Aschochyta pisi* Lib.

### Symptoms

This disease occurs worldwide and occasionally economically significant when cropping procedures and environmental conditions are favour for disease

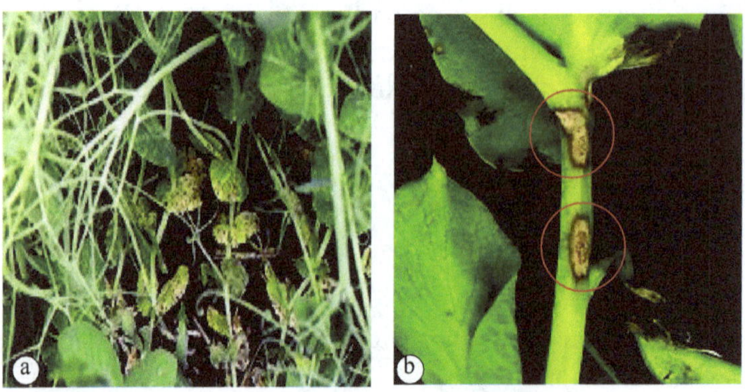

**Figure:** Pea Plants **(a)** and stem **(b)** with Ascrochyta blight

development. The small purple spots develop on the surface of the foliage, stems and pods. These brown spot are circular in shape, and crowded with minute black fruiting bodies of the fungus. The spots are marked with concentric zones with dark to black margins. The leaves may die in presence of many spots.

### Management

- All infected plant debris should be collected and destroyed.
- Spraying of Mancozeb @ 0.25% at 10-15 days intervals is recommended.
- Use of *Trichoderma viride* @ 4-5 kg/ha at the time of sowing

## 5. Downy Mildew

### Pathogen

*Pernospora pisi* (Berk.) de Bary.

### Symptoms

It appears early on young plants, in the form of downy growth on the under surface of leaves. As the disease advances the affected tissues turn brownish grey, dry up and the leaves drop off prematurely. If the young growing stems are systemically infected, they become distorted and is checked growth, with subsequent stunting of the entire plant. On pods, the disease is first noticed when they are green and flat. The patches are as pale green, more or less elliptical in shape, blotches on the sides, or more irregular, elongated lesions along the pods. The lesions gradually become bright brown in colour.

**Figure:** Downy mildew infected pea plant

### Management

- Since the primary inoculum comes from the soil, so sanitation and crop rotation will help in reducing the primary infection.
- Deep summer ploughing and eradication of weeds.
- Spraying of Redomyl @ 0.1%, Copper Oxychloride @ 0.3% and hexaconazole @ 0.1% at 10-15 days intervals are effective.

## 6. Anthracnose

### Pathogen

*Colletotrichum pisi*

### Symptoms

Pea anthracnose has been reported from several pea producing areas around the world, but is seen only rarely and is generally of minor importance. Leaf and stipule lesions are oval in shape, 2-8 mm in diameter, and have brown margins with grey tan centres. Lesions on the stem are long and similar in color to those on the leaves. Grey-tan pod lesions are round and sunken, with reddish brown borders. These are most dramatic when they are formed on immature pods, causing them to develop abnormally and show brown discolouration. Frequent wind-blown rain, high humidity and warm temperatures are most conducive for disease development.

### Management

- Follow 2-3 year crop rotations with non host crops.
- Alternate spray of Carbendazim @ 0.1% and Chlorothalonil @ 0.2% at periodical intervals are better solutions
- Grow disease resistant cultivars, if available.
- Avoid over irrigation during pod formations.

## 7. Rhizoctonia Seedling Blight

### Pathogen

*Rhizoctonia solani* (*Thanatephorus praticolus*)

### Symptoms

Rhizoctonia seedling blight, sometimes called rhizoctonia tip blight and stems rot of pea, occurs only occasionally and is generally of minor importance. The symptom of this disease is the death of the growing tip of very young pea seedlings. Rhizoctonia attacks up to ½ inch (1.2 cm) of the tip of the shoot just as it emerges through the soil and before leaves expand. Often one or two auxiliary shoots arise from the seed within a few days after the first. They may also become infected, or they may proceed to produce a normal plant. A tan to dark brown stem rot may occur near the ground surface area of pea seedlings.

### Management

- Rotate peas with crops known to be relatively poor hosts of this fungus, such as grain crops and corn.
- Seed should be treated with Thiram or Captan @ 2.5 – 3.0 g/kg seeds.
- Avoid irrigation during early stages of the crops.

## 8. Bacterial Blight

### Pathogen

*Pseudomonas syringae* pv. *phaseoli*

### Symptoms

The disease occurs worldwide. Symptoms appear on all above ground plant parts. They begin as small, water-soaked, oval spots. Individual leaf and stipule lesions enlarge to approximately 3 mm in diameter with an angular shape. Pod lesions are 6 mm or more in width, and stems lesions lengthen to resemble stem

**Figure:** Bacterial Blight affected pea plant

streaks. Lesions vary in colour, but on foliage they generally turn brown, shiny, and transluscent. Pods lesions, which may expand along either suture, slowly change from a water-soaked to greasy appearance with purplish-brown margins. Peas inside infected pods may develop brown discolorations. Multiple lesions often run together, forming large diseased areas which may cover large portions of infected plants. This may cause plants to look "blighted" and dry prematurely.

### Management

- Use disease free seeds/planting materials.
- Seed should be treated with Thiram or Captan @ 2.5 – 3.0 g/kg seeds.
- At least three year crop rotation with non-host crops should be followed.
- Grow resistant cultivars.

## 9. Mosaic

### Pathogen

Common Mosaic Virus

### Symptoms

Pea mosaic has been one of the common viral diseases in most of the important vegetable pea-growing areas of the world. Some mosaic outbreak results in significant economic loss, although ordinarily this is not the case. Distinct mottling of the foliage is the most conspicuous symptoms. The yellow tissue between the veins, leaving patches of normal green tissue scattered irregularly over the surfaces of both leaves and stipules. Vein clearing is not always apparent. If infected when young, plants become stunted. The upper leaves and stipules are wrinkled and twisted and auxillary buds often proliferate. Pods may be fewer and smaller then normal.

### Management

- Use resistant cultivars.
- Alternate sprays of insecticides when aphid populations reach the level of two aphids per plant.
- Rouging of infected plants.

## Bean/ Cow Pea

### 1. Leaf blight/Spot

#### Pathogen

*Cercospora cruenta* Sacc.

#### Symptoms

Symptoms appear as black, sooty and mouldy growth of fungus on all foliar parts. Initially the spores mass are olivaceous dark brown in colour and spread very fast in the field. Circular to irregular brown spots with reddish margin on leaves and all other plant parts are of common occurrence.

**Figure:** Leaf blight affected bean plant

#### Management

- Two to three sprays of Copper Oxychloride @ 0.3%, Hexaconazole @ 0.25% or Difenconazole @ 0.25% at weekly intervals.
- Collect the defoliated infected leaves and burn in field itself to reduce the inocula.
- Spray should be started immediately after disease initiation.

### 2. Sclerotinia blight

#### Pathogen

*Sclerotinia sclerotiorum* (Lib.) de Bary.

## Symptoms

It is a major disease of bean worldwide, particularly in cool, moist regions. Sclerotia overwinter in the soil as sclerotia and this fungus is favoured by cool, moist weather high humidity and long period of leaf wetness. Symptoms are observed as soft and white rotting of the tissues. Very soon white fungus growth is observed on the rotted portion.

Later on sclerotia embedded in white mycelium are formed on the infected portion as well as inner portion of pith and fruits. All pods and seeds may be attacked. Initial lesions are small, circular and rapidly increase in size. Affected tissues dry, turn brown and develop cottony mycelium.

**Figure:** Bean plant infected with Sclerotinia blight

## Management

- Seed cleaning is essential to remove the sclerotial population.
- Summer deep ploughing is highly desirable.
- All infected parts must be collected and burnt before drying up of the plant.
- Sanitation.
- Spraying of Ridomyl @ 0.1% or Mancozeb @ 0.25% at 10-15 days intervals is beneficial.
- Using *Trichoderma viride* @ 4-5 kg/ha before one week of sowing.
- Crop rotation with non hosts crops like, cereals and onion helps in less incidence of the disease.

## 3.Anthracnose

### Pathogen

*Colletotrichum lindemuthianum* Sacc. & Magn.

### Perfect stage

*Glomerella lindemuthianum*

### Symptoms

The typical symptoms develop in the form of angular red brown to black spots, sunken, crater like cankers on the pods. The central portion of the spot show pinkish masses of spores of the fungus in wet weather. Later on the sides of these spots appear as raised spots and develop on the cotyledons and stem of young seedlings which show one or two sunken, deep brown to black spots with ash coloured centre bearing pink spore masses. Severe infection results in defoliation and pod distortion. The pathogen survives in crop residues and in seed. It is disseminated within fields by rain, wind and favoured by cool temperatures, frequent rainfall and high humidity.

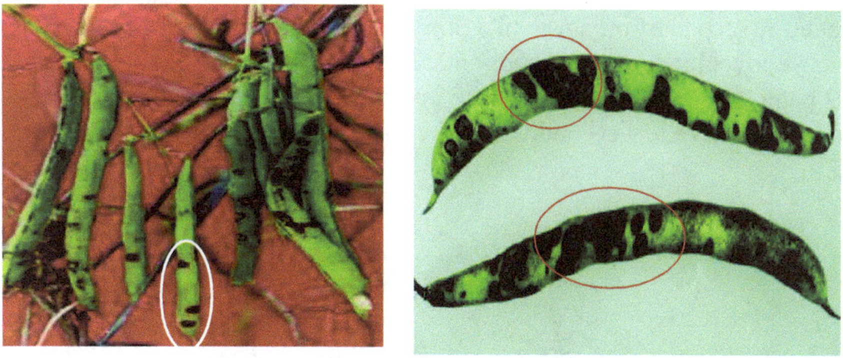

**Figure:** Bean fruits infected with Anthracnose

### Management

- Seed should be collected from healthy pods.
- Seed should be treated with Thiram @ 2.5 g/kg seed, Captan @ 2.5 g/kg or *Trichoderma viride* @ 4-5 g/kg seed before sowing.
- Two to three year crop rotation with non host crop such as cereal or maize should be followed.

- Infected plant debris should be collected and burnt.
- Spraying of Mancozeb @ 0.25%, Chlorothalonil @ 0.2% and Thiophanate methyl @ 0.2% at 10-15 days intervals just after the first appearance of the disease.
- Reduce humidity and high moisture contentwithin the fields by orienting rows in the direction of prevailing winds, avoiding excess irrigation after petal fall.

## 4. Downy mildew

### Pathogen

*Peronospora vicae* (Berk.) Gaum.

### Symptoms

Seedlings may get infected at emergence stage and die, although infected seeds usually fail to germinate. Systemically infected plants are light green with a silvery appearance. Local lesions begin as small, yellow blotches on the upper sides of leaves which are limited by leaf veins. The undersides of these lesions produce a grey to purple, fuzzy sporulation. Lesions become dry, brown and are invaded by secondary organism like Botrytis. Pods also get attacked and exhibits large, yellow blotches on the pods and white mycelial growth within the pods. This pathogen is seed and soil borne in nature. Disease development is most rapid under cool and wet conditions. The pathogen is disseminated by wind and rain.

**Figure:** Pods infected with downy mildew

### Management

- Rotate the crop with non-host crops to prevent the build up of soil borne inoculum.

- Plough under infected crop residue after harvest.
- Seed should be treated by Thiram or Captan @ 2.5-3.0 g/kg seed to protect emerging seedlings from infection.
- Applications of foliar fungicides have provided only limited management of this disease.
- Grow resistant cultivar, if available

## 5. Rust

### Pathogen

*Uromyces appendiculatus* (Pers.) de Bary

### Symptoms

The characteristic symptoms observed as minute, slightly raised pustules on all above ground plant parts. These pustules are distinct, yellow, circular sori consists of numerous uredospores which rupture the epidermis. They are often surrounded by a yellow halo. Severely infected leaves turn chlorotic, dry, and decline. Later on, dark red coloured teleutospores are formed on the leaves, pods and tendril. Ultimately the crops get dried.

**Figure:** Rust infected cowpea leaves

### Management

- Infected crop debris should be collected and burnt to reduce the inocula.
- Use of resistant cultivars.

- Spraying of Hexaconazole @ 0.1% or Propiconazole (Tilt) @ 0.1% at weekly intervals to reduce the inocula.
- Sulfer and Myclobutanil fungicides may provide some protection.
- Avoid over application of nitrogen and ensure adequate potassium fertilization.
- Rotate bean with non host crops
- Eliminate volunteer pea plants from the fields

## 6. Bacterial blight

### Pathogen

There are three major bacterial spot of bean:

### Bacterial Brown Spot

*Xanthomonas campestris* pv. *Phaseoli*

### Common blight

*Xanthomonas campestris* pv.*phaseoli*

### Halo blight

*Pseudomonas syringae* pv. *Phaseolicola*

### Symptoms

Symptoms of Bacterial brown spot first appear as small, circular, brown and necrotic spots, often surrounded by yellow zones. The lesions may enlarge, coalesce, and fall out giving leaves rugged appearance. Symptoms on pods are initially water soaked and circular, becoming brown and sunken. Distortion of pods can occur.

Symptoms of Common Blight initially appear as water soaked spots that enlarge, lose turgidity and become necrotic and bordered by a small zone of yellow tissue. Lesions may coalesce lending the plants to burnt appearance.

Symptoms of Halo Blight first appear on leaf undersides as small, watersoaked spots which develop in to numerous reddishbrown lesions with green to yellow halos which vary in size. In severe infections, it result in the bacteria becoming systemic and cause leaf yellowing and death. The pathogen enters the plants through natural openings or wounds and is spread by windblown rain, soil or plant debris.

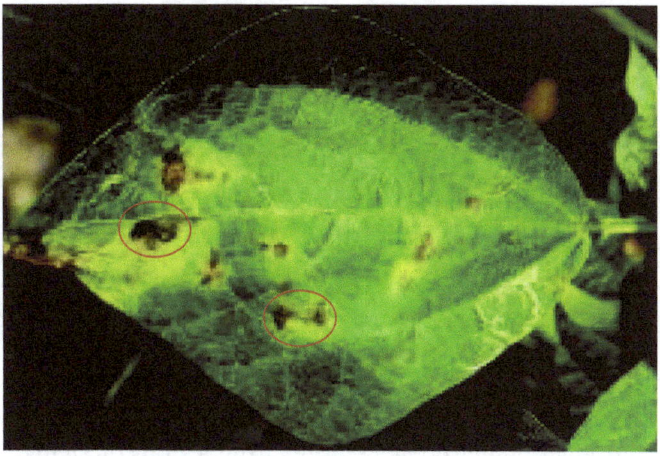

**Figure:** Bacterial blight infected bean leaf

## Management

- Only certified and disease free seeds should be planted.
- Use bean varieties with resistance or tolerance to the bacterial diseases.
- Rotate out beans with other non host crops for a minimum of two years.
- Streptomycin seed treatment can reduce surface contamination of the seed but well manage systemic infections of bacterial diseases.
- Weeds and volunteer beans should be eliminated from the fields.

## 7. Angular leaf spot

**Pathogen**

*Phaeoisariopsis griseola*

## Symptoms

All arial plant parts, including leaves petioles, stems and pods get infected, but symptoms are most recognizeable on leaves. Lesions on leaves usually appear as brown spots with a tan or silvery centre that are initially confined to tissue between major veins, which gives it an angular appearance. On some varieties yellow halo occasionally surrounding lesions appear and eventually the entire leaflet becomes yellow before senescing. Lesions can be observed on the under surface of the leaf and appear slightly more pale than those on the upper surface of the leaflets. Lesions on stems and petioles appear dark brown and elongated. Lesions on pods are circular, black and sunken and look similar to anthracnose.

The pathogen can survive between seasons on infected seed, which is one pathway of introduction in to fields. Spores from lesions on stems, leaves and pods as well as on crop debris help in perpetuation of this pathogen.

## 8. Bean Common Mosaic

### Pathogen

Virus

### Symptoms

Bean Common Mosaic is one of the most commonly prevalent viral diseases of French bean causing considerable losses in yield and quality. Yield losses up to 84 per cent have been reported due to the is disease. The symptoms of the disease appear as mixture of light yellow and green areas on the leaves. This is coupled with production of narrower leaves which roll upward, mottle, pucker and become malformed. Early infection leads to yellow coloured dwarf plants. Such plants usually fail to set pods or develop under sized pods. The pods on severely infected plants are usually of under size and contain fewer ovules than those produced on normal plants. They are occasionally covered with small, dark green spots and often mature very late. The seed may be smaller or malformed, or finally get aborted.

The disease is caused by Bean Common Mosaic Virus (BCMV) which belongs to poty virus group. The virus particles are filamentous, flexuous rods measuring 700-800 nm in length and 12-15 nm in diameter.

The virus is seed borne in a range of legumes. The virus can be transmitted in the non-persistent manner by several aphid species which do not normally colonize beans but transmit the virus as winged migrants especially *Acyrthiosiphon pisum, Aphis fabae* and *Myzus persicae*.

**Figure:** Bean leaf showing common mosaic

### Management

- Use of healthy and certified seed helps in checking the spread of the disease.
- Early planting, when the incidence of aphid vectors is low, gives better yields.
- Intercropping of maize with bean also reduces the incidence of this disease.

## 9. Root Rot and Web Blight

### Pathogen

*Rhizoctonia solani*

Root Rot and Web Blight are very serious disease of this crop throughout the world. In warm and wet conditions, it may assume serious proportions and cause significant losses. Losses from root rot vary greatly in different localities from year to year. Due to web blight yield losses of green pods at different stages of plant growth vary from 8.4 to 64.6 per cent.

### Symptoms

The symptoms of the disease appear in following two phases:

**Root rot:** Symptoms appear on roots and stem above and below the surface of the soil as reddish brown, sunken cankers. The lesions enlarge rapidly and an girdle the stem at the collar region, extending longitudinally downward to the roots leading to partial or complete rotting of the root system.

**Fig:** French bean roots showing root rot symptoms

**Web Blight:** The symptoms on leaves appear as small, circular, water soaked spots which later on become tan to brown in colour. These spots coalesce to form larger areas on the leaf blade. The pods are attacked at all stages of their growth. Dark brown, more or less circular, slightly zonate and definitely sunken spots appear on pods which in moist weather are covered with hyphae and sclerotia of the fungus. Tan brown to reddish brown discoloration is observed on infected seed.

## Management

- Always use disease free seeds.
- Infected plants should be collected and burnt.
- Seed treatment with Thiron and Captan @ 2.5g/kg of seed

# CHAPTER – 3

# Diseases of Solanaceous Crops

## Tomato

### 1. Damping off

#### Pathogen

*Pythium aphanidermatum*

*Fusarium* spp.

*Sclerotium* spp.

#### Symptoms

Damping off is the most important disease of tomato. Several seed and soil borne fungi can kill before the tender redical and plumule establish in the nursery bed. The pre emergence damping off and seed rotting is the usual cause of poor seedling stand. This type of damping off causes hidden loss but the post emergence damping off is more serious problem. Seedlings get infected near the soil line (collar region) after emergence and give wire stem symptoms. The tissue become soft, water soaked and weak causing the seedlings fall over and later on they die. The entire action is so fast that leaves remain green but they are unable to stand further.

**Figure:** Tomato seedling affected with domping off

### Management

- Soil solarization by the use of white transparent polythene in summer is one of the most effective management strategies of soil borne pathogens.
- Seed treatment with Thiram or Captan @ 2.5-3.0 g/kg seed or *Trichoderma viride* @ 6-10 g/kg seeds.
- Drenching of nursery beds by Thiram or Captan @ 2.5-3.0 g/litre water or Carbendazim @ 0.1% after seedling emergence also does work.
- Nursery bed should be properly drained and pulverized.

## 2. Early Blight

### Pathogen

*Alternaria solani* Ell. & Mart.

### Symptoms

Symptoms of this disease appear on the leaves as circular to angular, dark brown to black spots. The concentric rings often form leathery necrotic tissue. Usually a narrow chlorotic zone around the necrotic spots fades in to normal green. The infected leaves wither, droop and drop off. The stem and fruits are also affected. The infection is favoured by mild and rainy weather.

**Figure:** Early blight infected tomato plant

## Management

- Maintain field sanitation by plucking the lower leaves and burning of infected crop debris.
- Adopt summer ploughing to increase the desiccation of pathogen and infected plant parts.
- Spray of Mancozeb @ 2.5 g/litre and Chlorothalonil @ 0.2% at 10-15 days intervals are effective but spray must be started just after appearance of disease.
- Use of long duration crop rotation, eradicate weeds and elimination of volunteer plants and culling piles up in field help a lot.
- Healthy seeds should be collected from diseased free fruits.

## 3. Late blight

### Pathogen

*Phytophthora infestans* de Bary

### Symptoms

Symptoms can appear on any above ground part of the plant at any stage of the crop growth. Symptoms on leaves, twigs and fruits are more common. Blight appears on foliage as light pale green water soaked dead areas. The lesions are wet in morning and dried, shriveled in day night. It enlarges rapidly until entire leaves are killed and exposing all the fruits for infection. Green and ripe fruits affected by the blight appeares as greenish brown and internal tissues become spongy with odour.

**Figure:** Late blight affected tomato plant

## Management

- Infected plant debris and fruits must be collected from the field and burnt.
- Spray of Mancozeb @ 0.25% provides good control of the disease.
- One spray of Sixer (Metalaxyl + Mancozeb) @ 0.2% is very effective when applied within two spray of infection but repetitive spray should not be given.
- Spray of Azoxystrobin @ 0.1% at the first sign of disease appearance and rotate it with Chlorothalonil @ 0.2% after 7-15 days if necessary.
- Grow resistant cultivars like TRB-1 and TRB-2.
- Stacking of plant reduces all *Phytophthora* disease of tomato.

## 4. Pseudocercospora leaf blight

### Pathogen

*Pseudocercospora fuligena* (Lev.) Speg.

### Symptoms

Black to olivacious black and sooty growth of fungus are observed on the upper surface of the leaves. The corresponding lower portion of leaves is light yellow to tan coloured. Leaves defoliated soon after infection. Symptoms are also observe on petioles, calyx and stem.

**Figure:** Tomato plants/leaves affected with pseudocercospord leaf spot

### Management

- One to two sprays of Hexaconazole and Tridemephon @ 0.05% at 10-15 days intervals help in suppressing the disease.
- Infected plants parts should be collected and burnt.

## 5. Fruit rots

### Pathogen

*Colletotrichum gloeosporoides* Penz.

### Symptoms

Several fruit rotting pathogens particularly in kharif season affect tomato and cause severe losses. *Pythium* and *Phytophthora* affect green fruit while *Rhizoctonia, colletotrichum* and *Sclerotium* affect mature fruits. Symptoms of *rhizoctonia* fruit rot is rhythmic rotting in nature of fruits while coming in contact with in the soil. Rotting is immediately followed by cracking and fungal growth. *Sclerotium* fruit rot is very clear due to white fungal mycelium and mustard grain like sclerotia over fruits. Generally lower portion of fruits are infected which are coming in soil contact.

### Management

- Avoid the contact of fruit with soil by stacking of plant.
- Green manuring followed by soil application of *Trichoderma viride* @ 5.0 kg/ha does wonder. In soil it is very effective in checking most of the fruit rot disease.

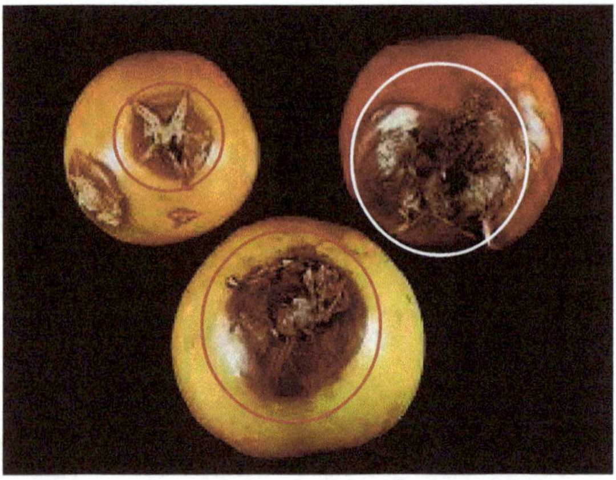

**Fig:** Tomato fruits infected with Fruit rot

- Collect affected fruit and burn them to reduce primary inoculum.
- Provide proper drainage in the field.
- Rotate tomato with non-solanaceous crops for at least one year
- Stacking and mulching can also reduce losses caused by anthracnose.

## 6. Fusarial Wilt in Tomato

### Pathogen

*Fusarium oxysporum f. sp. lycopersici* Brushi.

### Symptoms

Clearing of veinlets, dropping of petioles, yellowing of lower leaves and wilting and dying of the plant are common symptoms The pathogen is soil borne in nature and drainage water transmits the disease in soil.

### Management

- Two to three year crop rotation with non host crops in advisable
- Soil solarization before nursery raising is good
- Use of *Trichoderma viride* @ 4-5 kg/ha and vermicompost @ 2-3 t/ha at the time of nursery raising.
- Sanitation and use of disease resistant cultivars.
- Green manuring followed by application of *Trichoderma* @ 5 kg/ha with in a week of ploughing.

**Figure:** Wilt affected tomato plant

- Application of recommended fungicides at flowering stage of the crop.

## 7. Collar Rot

### Pathogen

*Sclerotium rolfsii* Sacc.

### Symptoms

The initial symptoms are observed as necrosis of soft tissues of bark of the stem near the collar region. White pathogens growth is clearly visible on the affected portion and just below the soil surface. Progressive dropping and yellowing or wilting of the entire plant and sometime plant collapse. Disease is soil borne where in pathogen survives in the form of sclerotia.

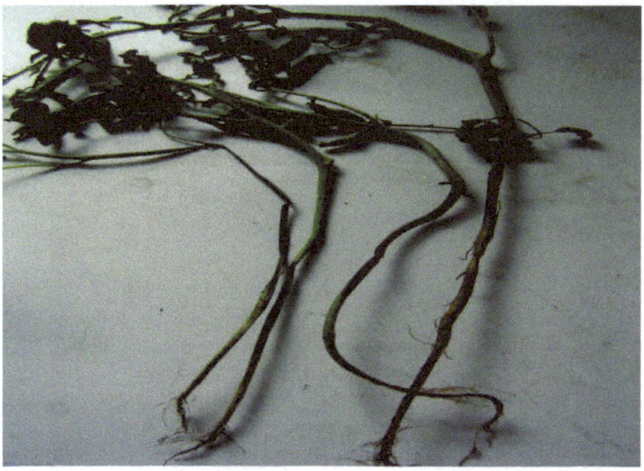

**Figure:** Collar rot infected tomato plants

### Management

- Remove all the weeds from the field.
- Irrigate the soil in summer and plough the field three times to destroy germinating sclerotia.
- Avoid high density planting.
- Drenching with Captan @ 0.25%, Carbendazim @ 0.1% near the coller region is effective to check the infection.
- Crop rotation with cereals, corn, sorghum and millets are good option.
- Green manuring followed by application of *Trichoderma* @ 5 kg/ha with in a week of ploughing help a lot.
- Drenching with Copper Oxycholoride @ 0.3% near the collar region followed by Carbendazim @ 0.1%

## 8. Sclerotium Fruit Rot

### Pathogen

*Sclerotinia sclerotiorum* (Lib.) de Bary

### Symptoms

If the soil is wet, mass of white mycelium of the fungus can be seen on the diseased portions. Mustard seed like sclerotia of the fungus can be seen on the mycelial mats. Fruits in contact with the soil can be attacked first. Water soaked, soft, star shaped lesions appear on fruits and the affected fruits collapse within 3-4 days. The lesion cavity shows white mycelium with sclerotia.

**Fig:** Black scleratia on Tomato fruits

### Management

- Solarization of the soil before nursery raising
- Use *Trichoderma viride*, *Pseudomonas fluorescence* and *B. bassiana* in soil before transplanting.
- Crop rotation with non susceptible/non host crops is desirable.
- Summer ploughing with proper sanitation reduces the sclerotial population in soil.
- Seedling dip in *Trichoderma* sp @ 1% for 10 minutes before transplanting is effective against this disease.

## 9. Bacterial Wilt

### Pathogen

*Pseudomonas solanacearum*

### Symptoms

The initial symptoms appear an a few young leaves as these become flaccid due to lack of moisture. When the infected stem is cut horizontally, droplets of yellowish, sticky ooze are visible on the cut surface. Wilted plants eventually show rotting of roots.

**Fig:** Bacterial wilt affected tomato plants

### Management

- Two to three year crop rotation with cereals.
- Use of resistant cultivars like, Arka Alok, Arka Vardhan, Utkal Pallvai (BT1), Utkal Kumari (BT 10) and BWR-1.

## 10. Bacterial Spot

### Pathogen

*Xanthomonas campestris* pv. *vesicatoria*

### Symptoms

The disease is prevalent throughout the country. Symptoms of small, dark spots appear on leaflets and stems of seedlings transplanted crops. Numerous spots coalesce and cause severe burning of the entire leaves. This disease is more severe in *kharif* season and more conspicuous and damaging phase are the infection of immature fruits. Small water soaked spots appear on exposed surfaces of the green turning red fruits.

**Figure:** Tomato plant infected with Bacterial spot

### Management

- Soil solarization in nursery beds to avoid seedling infection.
- Healthy and uninfected seeds should be collected from disease free plants.
- Alternate spray of Streptocycline @ 150-200 ppm and spray of Copper Oxychloride @ 0.3% at 15 days intervals are quite useful.

## 11. Leaf Curl

### Pathogen

Virus (*Gemini Virus*)

### Symptoms

Leaves show downward rolling, curling, crinkling, twisting and chlorosis. The newly formed leaves show chlorosis. The older leaves become leathery and brittle. Infected plants look pale and produce more lateral branches giving a bushy appearance. This virus is transmitted by White fly (*Bemisia tabaci*) as well as through mechanical injury also.

### Management

- Remove infected plants soon after noticing infection at initial stage of growth.
- Collect healthy seeds from disease free fruits/plants.
- Use barrier crop of taller nature of non-host crops like, Maize, Bajra and Sorghum.
- Periodical spray of recommended systemic insecticides up to flower setting is advisable

**Figure:** Leaf curl affected tomato plant

## 12. Mosaic

### Pathogen

Virus (*Gemini Virus*)

### Symptoms

In early stages, symptoms lead to stunted chlorotic plants and leaves showing curling, crinkling and downward rolling depending on the leaves upon the virus strains. In severe cases, plant growth remains stunted and ultimately die.

**Figure: Mosiac of Tomato (www.avrdc.org).bmp**

### Management

- Always used treated seeds
- Use of resistant/tolerant cultivars.
- Removal of infected plants at initial stage
- Collected seeds from healthy plants.
- Alternate spray of Imidacloprid @ 0.4% and Monocrotophos @ 0.2% at 10-15 days intervals.

## 13. Root- Knot Nematode

### Pathogen

*Meloidogyne incognita, M. javanica*

### Symptoms

The first symptom of the disease is stunting and poor fruiting. If the affected plants are pulled out and examined, the roots will be found severely knotted. In some cases the plants may show chlorosis but rarely do the plants die from nematode infection. The most characteristic symptoms is the appearance of swelling called galls or root-knots all over the root system

## Management

- Soil solarization during nursery raising.
- Crop rotation for 2-3 year with cereal crops.
- Grow resistant cultivars (SL-120, Hissar lalit etc).
- Useing Nemagon @12-15 kg/ha or Phorate @ 10 kg/ha before one week of transplanting
- Using of biocontrol agents like, *Paecelomyces*, *Metarhizium* and *Baeveria bassiana* and botanical like, Neem cake at the time of transplanting

**Figure: (a)** Root knot infected tomato root **(b)** healthy tomato root

## Capsicum/Pepper

### 1. Powdery Mildew

#### Pathogen

*Leveillula taurica* (Lev.) Arnaud.

#### Symptoms

Symptoms appear as small, circular, whitish spot on both the surface of the leaves. In later stages entire leaf surface is covered with dirty white growth.

The infection starts from the older leaves but gradually covers the entire plant. In such cases, the lamina develops incompletely or partial chlorosis and ultimately plant dries up. In later, the mildews mass turns grey and develops numerous minute black bodies distributed all over the surface.

**Figure:** Whitish powdery mass on capsicum leaves/fruit

### Management

- Before flowering spraying of Sulpher, Wettable sulpher or Karathane @ 0.2% is effective.
- Grow resistant cultivars
- Spraying of Carbendazim @ 0.1% at 10-15 days intervals most beneficial

## 2. Anthracnose/Fruit rot

### Pathogen

*Colletotrichum capsici* (syd.) Butler & Bisby

### Symptoms

Symptoms appear mostly on ripened fruits. The spots are usually circular and sunken in the shape with black margins. The sunken spot is covered with a pinkish mass of fungal spores. As disease advances the spots spread forming concentric markings with dark fructifications representing the fungal acervuli. The fruits with many spots drop off prematurely resulting in heavy loss in yield. The fungus may also attack the fruit stalk and spread along the stem causing dieback symptom.

**Figure:** Capsicum fruit infected with Anthracnose disease

### Management

- Seed treatment with Thiram + Captan 1:1 (0.3% concentration) is recommended.
- Alternate spray of bavistine @ 0.1%, Difolatan @ 0.2% and Mancozeb @ 0.25 is effective for disease management.
- Seed treatment with *Trichoderma viride* @ 6-10 g/kg seed is effective for disease prevention.

## 3. Bacterial Wilt

### Pathogen

*Pseudomonas solanacearum*

### Symptoms

Symptoms of the disease are clearing of the veinlets and chlorosis of the leaf. Soon the petiole and leaves droop and wilt. The younger leaves may die in succession and the entire plant may wilt and die in the course of a few days.

### Management

- Sanitation and crop rotation
- Use of resistant/tolerant cultivars

**Figure:** Wilt affected capsicum plant

## 4. Bacterial Spot

### Pathogen

*Xanthomonas compestris* pv. *vesicatoria*

### Symptoms

The first symptom (spots) appears on leaves, the spots are generally brown, circular, and water soaked. Bacterial spot lesions do not have concentric zones or a prominent lesions . When conditions are optimal for disease development, spots can coalesce to form long and dark streaks. A general yellowing may appear on foliage with many lesions giving the plants a scorched appearance. Symptoms on ripe fruit may be scab like or sunken area. The bacteria are able to survive on tomato volunteer plants and can overwinter on diseased plant debris. Seed is an important mechanism for survival and dissemination of the pathogens.

**Figure:** Capsicum leaf infected with bacterial spot

### Management

- Rotate fields to avoid carryover on volunteers or crop residues.
- Keep field free from volunteer's plants, weeds and cull piles.
- Apply appropriate bactericides or combination of bactericides.
- Streptocycline sulfate @ 200 ppm can be used only up to transplanting stage.

## 5. Leaf Curl Complex

### Pathogen

*Virus*

### Symptoms

Leaf curl complex is mainly due to CMV and Gemini virus as well as mite and thrips which act as vectors. Reduction in size of leaves, shortening of veins, puckering, mottling of the leaves, stunting and bushy appearance of the plant are some of the symptoms of viral infection. Downward rolling, narrowing and shortening of leaves in mite infestation.

### Management

- Initial roughing of infected plants soon after infection and burn it.
- Root dipping of the seedling in Imidacloprid solution @ 4-5 ml/litre of water for one hour before transplanting.
- Nursery should be grown in nylon net to check the vector infestation.
- Barrier crop of taller nature of non-host crop like Maize, Bajra and Sorghum should be planted.

**Figure:** Leaf curl affected capsicum plant/leaves

- Alternate spray of Dicofal @ 0.25% with wettable Sulpher @ 0.2% and 1-2 spray of systemic insecticide.

## 6. Root-Knot of Capsicum

### Pathogen

*Meloidogyne hapla*

### Symptoms

Symptoms of root-knot nematode are characteristic of root damage. Stunting, wilting, chlorosis and reduced fruit yield and size. Plants fail to respond to fertilizer applications. Root-knot nematode enters roots through root hairs and become sedentary (endoparasites). They induce giant cells to form in the root system which are apparent as galls or knots. Galls may also be invaded by secondary organisms that cause further root decay.

### Management

- Deep summer ploughing reduce the nematode population in soil.
- Crop rotation with wheat, sorghum, corn or legumes may be effective
- Soil solarization may reduce populations in the upper layers of the soil.
- Chemicals or fumigant nematicides can be used for high value crops.
- Use Nematode resistant cultivars.

## E. Chilli

## 1. Leaf Spot/Blight

### Pathogen

*Alternaria solani* Ell. & Mart.

### Symptoms

Circular to oblong spots with light grey to white centres and dark brown margins. Small spots coalesce to each other in severe case and lead to defoliation. In moist condition, sporulations of both pathogens are observed on the spots.

### Management

- Field Sanitation by plucking the lower leaves and burning of infected plant debris.

**Figure:** Concentric spot on chilli fruit

- Summer ploughing.
- Foliar spray of Chlorothalonil @ 0.2% and Mancozeb @ 0.25% at 10-15 days intervals are good.
- Healthy seed should be collected from disease free fruits.
- Crop rotation with non-solanaceous crops to reduce the inoculum.

## 2. Powdery Mildew

### Pathogen

*Leveillula taurica* (Lev.) Arnaud.

### Symptoms

Small, circular, whitish spot in lower surface of the leaves. In later stage the entire leaves surface is covered by dirty-white powdery growth. The infection starts from the older leaves but gradually covers the entire plants.

### Management

- One to two spray of Propiconazole @ 0.1% or Sulpher @ 0.2% just after appearance of the disease.
- Use of resistant cultivars.
- Avoid over head irrigations.

## 3. Fruit Rot and Die-back

### Pathogen

*Colletotrichum capsici* (Syd.) Butler & Bisby

### Symptoms

Symptoms on ripe fruits are more common than green fruit. Small, black, irregular, sunken lesions were appeare on mature fruits. Most oftenly they coalesce and become papery on green fruits. Acervuli formation start from the centre of lesion and moved towards periphery on red fruits. The infected parts turn dark and become depressed or wrinkled and ultimately the fruit shrivels and dries up.

### Management

- Seed should be collected from healthy fruits.
- Seed should be treated with Thiram @ 0.25% or Carbendazim @ 0.1% before sowing.
- Spray Carbendazim @ 0.1% or Copper Oxychloride @ 0.3% at the time of flowering.
- Grow resistant Cultivars like Punjab Sukh etc.
- Seedling should be treated with Carbendazim @ 0.15 before transplanting.

**Figure:** Fruit rot affected chilli fruit

## 4. White/Sclerotinia Rot

### Pathogen

*Sclerotinia sclerotiorum* (Lib.) de Bary

### Symptoms

The symptoms appear as water soaked rotting of curd, petiole, stalk and stump region of the leaves. After infection, growth of white mycelium is observed all over the infected portion. Seed production crops are severely affected by this disease and entire inflorescence collapsed. In later stages the total rotted portion is converted in to compact mat followed by hard black sclerotial body.

### Management

- Alternate spray of Mancozeb @ 0.25% with Carbendazim @ 0.1% using sticker at flowering stage.
- All the infected plant parts should be collected and burnt.
- Deep summer ploughing decreases inoculum.

## 5. Leaf Curl

### Pathogen

Virus

### Symptoms

The characteristic symptoms of the virus infection are vein clearing, curling, blistering, puckering and heavy crinkling of the leaves with chlorosis and enations. Infected plants are stunted and often bushy in appearance.

### Management

- Selection of seed from healthy fruits.
- Nursery should be grown under controlled condition.

**Fig:** Curling and crinkling chilli plant/leaves

## Brinjal

### 1. Leaf spot

#### Pathogen

*Alternaria melongenae* Rangaswami & Sambandam

#### Symptoms

Brown and irregular spot are produced on upper surface of the leaves. Concentric rings are present in the spots. Several spots may coalesce to form bigger necrotic patches. The leaves may drop off due to severe infection. The infection on the fruit causes large deep sunken spots. Infected fruit turn yellow and drop prematurely.

**Figure:** Leaf spot infected brinjal leaves

## Management

- Proper field sanitation is essential.
- Spray of Mancozeb @ 2.5 g/litre of water and Sixer (Metalaxyl + Mancozeb) @ 0.2% are effective against this disease.
- Use of healthy seeds from disease free plants.
- 2-3 year crop rotations with non solanaceous crops are required.

# CHAPTER – 4

# Diseases of Cole Crops

## Cauliflower, Cabbage and Knol khol

### 1. Leaf spot

#### Pathogen

*Alternaria brassicae* (Berk.) Sacc.

#### Symptoms

Alternaria leaf spot usually appear in early stage in cauliflower while at later stage symptoms in cabbage appear as circular light brown spot in leaves. Concentric rings are clearly visible on the spots. In older plants on leaves, petioles and stem, brown to black circular to slightly elongated spots appear.

**Figure:** Leaf spot affected cauliflower leaf

## Management

- Use of disease free seeds from healthy plants.
- It can be effectively managed by detaching all the infected lower leaves in morning and then burning.
- Spray Chlorothalonil @ 0.2% and Mancozeb @ 0.25% alongwith sticker at the first sign of disease.
- Practise long rotations with nonbrassicae crops.
- Avoid excess irrigation during head development.

## 2. Black Leg

### Pathogen

*Phoma lingam* (Fr.) Desm.

### Symptoms

Symptoms start on young seedlings as blueish black discoloration on the stem. This lesion grows in to a sunken brown lesion with a purple black border. Small black spots may be seen in the centre of the lesion. These are spore producing structures and release a coil of pinkish in wet conditions. The stem lesion extends in to the soil and may cause discolouration and death of plant roots. The lesions eventually engirdle the entire stem, causing the plant to wilt and lodge. The fungus is transmitted though infected seed and occasionally through air borne spores.

### Management

- Use healthy and disease free seeds.
- Rotate with non cruciferous plants for three years.
- Also avoid sprinkler irrigation
- Dispose of the crop residue at the end of the season.

## 3. Sclerotinia Rot

### Pathogen

*Sclerotinia sclerotiorum* (Lib.) de Bary

### Symptoms

This disease is more common in cole crops especially in the seed production crops. Symptoms appear as water soaked rotting of curd, petiole, stalk and stump region of the leaves. After infection, growth of white mycelium is observed all over the infected portion. The seed production crop is severely affected by the disease and entire inflorescence gets collapsed. The mycelium develops honey dew stage after colonization of the tissue. Later on, entire rotted portion is converted in to compact mycelial mat followed by development of hard black sclerotial body.

### Management

- Cut the infected curds, leaves along with some healthy portion in morning and carefully collect in polythene to avoid falling of sclerotia in the field.
- Foliar spray of Carbendazim @ 0.1% and Mancozeb @ 0.25% along with sticker at flowering stage.
- All the infected plant parts should be collected and burnt.
- Deep summer ploughing reduces the inoculums in soil.

## 4. Club Root

### Pathogen

*Plasmodiophora brassicae* Woron.

### Symptoms

Club root is a serious disease that occurs worldwide. The infection process begins when the resting spores germinate and enter the plants through root hairs or wounds. The development of this disease is favoured by warm temperatures, high soil moisture and an acidic pH. The initial symptoms of club root are difficult to detect. Later symptoms include pale yellow leaves and tendency to wilt during hot and sunny days. Young plants may be killed by the disease within a short time after infection, whereas older plants may survive but fail to produce marketable heads.

**Figure:** Club root infected cauliflower leaf and plant

### Management

- Avoiding planting any cruciferous plant in the same fields for long time periods.
- Always use disease free seeds.
- Follow crop rotation with non-host.

## 5. Black Rot

### Pathogen

*Xanthomonas campestris* pv *campestris*

### Symptoms

Black rot, which is distributed worldwide, is the most serious disease affecting cabbage and cauliflower. This bacterial pathogen can infect all crucifers including allied crops and related weeds. A disease symptom usually appears first on the edges of the leaves. The bacteria enter through leaf pores and wounds, eventually producing a yellow V-shaped area on the margin of the leaf, with the point of the V following a vein towards the leaf petiole. The veins within infected areas eventually turn black, and the leaf tissue becomes dry, brown and brittle. The bacteria then move systemically throughout the plant. When infected petioles are cut open, veins are clearly black and may ooze a stricky yellow bacterial slime. Ultimately plants become stunted, wilt and die.

### Management

- Use only certified disease free seed.
- Rotate with non cruciferous plants for three years.
- Manage all weeds, especially cruciferous weeds.
- If disease shows up in seedlings, destroy all infected seedlings and closely monitor neighbouring seedlings for symptoms.

## 2. Phomopsis Blight/Fruit Rot

### Pathogen

*Phomopsis vexans* (Sacc. & Syd.) Harter.

### Symptoms

The pathogen affects the plants from seedling stage to maturity. Leaves show clearly defined circular, light brown spots mostly in lower leaves. Centrr of the spot is lighter in colour. Old spot are having numerous black dots like pycnidia. These spots are papery, often crack and some time form shot hole. Stem lesions start from basal part near nodal region as grey dry rot with constriction. This leads to drying of few twigs or partial wilting of the some portion of plant. Old and dried twigs show several dots like erumpent pycnidia on the bark. Pale to

**Fig:** Phompsis blight infected brinjal fruit

light brown sunken spots develop on the old fruits. Fruit rot is soft, spongy and with rhythmic growth. Later on the infected fruit is completely rotted and mummified due to dry rot. Fungus perpetuates in the soil and lives on affected tissues.

### Management

- Seed should be collected from healthy fruits and disease free field.
- Seed treatment with Carbendazim @ 0.25% and *Trichoderma viride* @ 6-10 g/kg seed is essential to eliminate the infection.
- At least 2-3 years crop rotation is required with cereals.
- Spray of Carbendazim @ 0.1% after 15 days of transplanting and during flower setting serves well.
- Infected crop residue should be collected and burnt.
- Grow resistant cultivars like Pusa Vaibhav.
- Two to three year crop rotation with cereals are good for suppress of the pathogen.

## 3. Sclerotinia rot

### Pathogen

*Sclerotinia sclerotiorum* (Lib.) de Bary

### Symptoms

The infection is mainly on the stem and branches. At the point of infection a dry, discoloured spot develops which gradually on girdles the stem and also progresses up and down. As a result of severe infection the portion of the plant beyond the infection wilts. White mycelium develops honey dew stage soon after colonization. Later the entire plants may wilt in case of infection at the base of the stem. The pathogen is soil borne as well as concomitant with seed mixture.

### Management

- Carryout deep summer ploughing in such a way that surface residues is buried deep.
- Green manuring of Sunhemp in June and July followed by soil application of *Trichoderma viride* @ 4-5 kg/ha does excellent.

- Crop debris and infected plant parts should be carefully cut along with some healthy parts and collected from the field in morning. These should be burnt well before shattering and falling of scelerotia take place in the field.
- Alternate spray of Mancozeb @ 0.25% and Sixer (Metalaxyl + Mancozeb) @ 2.0% at 10-15 days intervals is effective.
- Follow suitable crop rotation with onion, beet, spinach or maize.
- Dense planting should be avoided to reduce the shade and humidity in plant canopy.

## 4. Verticillium Wilt

### Pathogen

*Verticillium dahliae* Nees & Wallr.

*Verticillium albo-atrum* Reink. & Berthhold

### Symptoms

Symptoms are subtle and may be confused with insufficient moisture availability to plants or other vascular wilts. Yellowing of lower leaves followed by wilting is the first sign of disease. Lesions are surrounded by a large, irregular area of yellowing due to a systemic leaf toxins produced by the fungus. Leaf necrosis is followed by wilting, stunting, and plant death. When the stems of infected plants are cut lengthwise, the vascular tissues exhibit brown discolouration.

**Figure:** Verticillium wilt affected brinjal plant **(a)** and stem **(b)**

### Management

- Soil solarization during nursery raising.
- Remove and destroy infected crop debris and throw them away from production fields.
- Crop rotations with non susceptible hosts are best options.

## 5. Bacterial Wilt

### Pathogen

*Ralstonia solanacearum*

### Symptoms

The disease is prevalent in warmer climates having high humidity under tropical and sub-tropical conditions. The characteristic symptom of the disease in sudden wilting of the plants. When cut across the infected roots of the plants, whitish bacterial ooze can be seen under low power of microscope. Vascular bundles are found to be discoloured dark brown. Pathogen can be attack any stage of the crop except in nursery

### Management

- Application of biocontrol agent like *Pseudomonas fluorescence* can be effective against this disease.
- Spraying of antibiotics like Streptomycin, Tetracycline @ 0.1% at periodical intervals etc.
- Grow resistant cultivars like, Arka Nilkuntha, Arka Keshev, Arka Nidhi, Swarna Prabha, BWR-12, Utkal Tarini, Utkal Madhuri and Annamalai. etc.
- Follow long crop rotation with cereals and non-solanaceous hosts.
- Avoid overhead irrigations during entire cropping period.

## 6. Little leaf

### Pathogen

*Mycoplasma*

### Symptoms

The characteristic symptoms of the disease are the production of very short/

little leaves by the affected plants. Numerous leaves aggregate in each branch giving bushy appearance and shortening of internodes. Auxiliary buds enlarge but petioles and leaves remain small, narrow, pale green, soft, smooth leaves and small fruit. Later stage of infection leads to complete sterility of the plant. Infected plants do not bear any fruit and if fruiting occurs, the fruits are very hard. The disease is caused by Mycoplasma Like Organism (MLO) and it is transmitted through leaf hoppers.

**Figure:** Mycoplasma infected brinjal plant

# CHAPTER – 5

# Diseases of Cucurbit Crops

## 1. Damping off

### Pathogen

*Pythium* spp
*Phytophthora* spp
*Fusarium* spp
*Sclerotium* spp

### Symptoms

It is a term used to describe the damage caused by various fungi to the seed and emerging seedling in the reduced plant stand and sometimes it is so serious that resowing becomes inevitable. Major symptoms are the failure of the seedlings decay or sudden death of the emerged seedlings roots and lower stems of emerged seedlings become soft and watery and may be reddish brown in colour.

### Management

- Prepare raised bed nursery about 10 cm above ground level for good drainage to avoid damping off.
- Follow soil solarisation for two to three weeks using 0.45 mm thick polythene (LLDPE) sheet. Suffecient moisture should be present in the

soil for solarization to taken place.

## 2. Anthracnose

### Pathogen

*Colletotrichum laginarium* Corda.

*Colletotrichum orbaculare*

### Symptoms

Disease symptoms on all above ground plant parts start from cotyledon leaf to fruits. Symptoms on leaves are observed as water soaked, small yellow spots that enlarge and turn to brown. The necrotic portion get dry and shatters. Elongated water soaked, sunken lesions appear on stem.

**Fig:** Anthracnose infected truck melon fruit

### Management

- Collect seeds from healthy fruits.
- Seeds must be treated with Thiram or Captan @ 2.5-3.0 g/kg seeds.
- Ensure field sanitation by burning of plant debris
- Spray of Chlorothalonil @ 0.2% and Mancozeb @ 0.25% give good management of the disease but spray must be started before infection.
- Avoid over head irrigation during entire cropping period.

**Figure:** Bitter gourd infected with Anthracnose/Fruit Rot

**Figure:** Bottle gourd infected with Anthracnose

## 2. Powdery mildew

### Pathogen

*Erysiphe cichoracearum* DC.

### Symptoms

Symptoms appear on all foliar parts, mostly confined to the upper surface, but also found on lower surface and the stem as white to dull white powdery growth.

This white growth quickly covers most of the leaf surface and leads to heavy reduction in photosynthesis area. Plant may wither and die.

**Figure:** Cucumber plant/leaf infected with Powdery Mildew

## Management

- Spray of any sulpher fungicides @ 0.2% and Penconazole @ 0.05% is effective against this disease.
- Proper drainage is required.
- Grow resistant varieties.

## 3. Leaf spot

### Pathogen

*Cercospora citrullina* cooke.

### Symptoms

The first symptom is appearance of minute, water soaked lesions on the leaf blade. These lesions soon enlarge and develop in to dark brown to greyish spots, mostly angular in shape, bound by the leaf veins. A few to over a hundred spots may get developed on a leaf blade, depending on the severity of infection. In severe infection get entire plant may wither and die.

**Figure:** Leaf spot affected pumpkin plant/leaf

### Management

- Alternate spray of Mancozeb @ 0.25% and Chlorothalonil @ 0.2% at 10-15 day intervals
- Seed treatments with Thiram, Captan @ 2.5-3.0 g/kg seed and *Trichoderma viride* @ 6-10 g/kg seed.
- Field sanitation, selection of healthy seeds and crop rotation reduce disease incidence.

## 4. Downy Mildew

### Pathogen

*Pseudopernospora cubensis*

### Symptoms

The first symptom appears on the leaves and resembles those of mosaic mottling. The pale green areas get separated by islands of darker green hols. Spots become well defined. They are angular, yellow coloured and often restricted by veins on the upper surface. On the lower side of these spots a purplish downy growth appears in moist weather. The entire leaf dies quickly. In later stage the entire plant is wilted or weakened. Infection occurs more readily on lower surface of leaves than on upper surface.

**Figure:** Downy mildew infected cucumber plant

### Management

- Spraying of mancozeb @ 0.25%, Chlorothalonil @ 0.2% or Copper Oxychloride @ 0.3% at 10-15 days intervals.
- Remove infected vines and burn it them
- Grow resistant cultivars.
- Bower system of Taining reduces the disease incidence.

## 5. Wilt of Cucurbits

### Pathogen

*Fusarium oxysporum f. sp cucurbitae*

### Symptoms

The infected plant shows sudden wilting during the mid season of the crops. The disease can be distinguished from the vascular wilt by the dark brown cortical soft decay at the base of the stem. The underground parts are found disintegrated. In humid weather the stem base may be seen covered with fungal growth.

### Management

- Treat the seed with Thiram or Captan @ 2.5 g/kg seed or Carbendazim @ 1 g/kg seed.
- Proper drainage is essential.
- Crop rotation with non-cucurbitaceous crops is need of hour
- Summer ploughing is a must.

**Figure:** ?

## 6. Mosaic

### Pathogen

Virus

### Symptoms

Disease is characterized by the formation of streaks in the interveinal regions of the leaves, which enlarge to form characteristic green veinbanding. Young leaves are usually distorted with wavy or irregular margins and wrinkled surface with green blisters. The entire plant may look stunted and flower little or none fruiting may be very much reduced both in number and size.

**Figure:** Cucumber plant infected with mosaic

## Management

- Alternate spray of systemic insecticides like Imidaclorpid (Jumbo) @ 0.4 ml/litre, Monocrotophos @ 2.5 ml/litre or Endosulphan @ 2.5 ml/litre at 10-15 days intervals.
- Removal of infected plants at initial stage.
- Grow resistant cultivars.
- Use of virus free seeds

## Root knot nematode

## Pathogen

*Meloidogyne incognita*

*M. javanica*

## Symptoms

Typical symptoms of nematode injury can involve both above ground and below ground plant parts. Foliar symptoms of nematode infestation of roots generally involve stunting and general unthriftiness, premature wilting and slow recovery to improved soil moisture conditions, leaf chlorosis (yellowing) and other symptoms characteristic of nutrient deficiency. An increased rate of ethylene production, thought to be largely responsible for symptom expression in tomato, has been shown to be closely associated with root-knot nematode root infection and gall formation. Plants exhibiting stunted or decline symptoms usually in patches of no uniform growth rather than as an overall decline of plants within entire field

**Figure:** Galled roots (left) infected by *Meloidogyne* sp

**Figure:** Root-knot nematode (*Meloidogyne* sp.) induced galling of watermelon roots

## Management

- Soil solarization during nursery raising.
- Crop rotation for 2-3 year with cereal crops.
- Grow resistant cultivars (SL-120, Hissar Lalit etc).

- Using Nemagon @12-15 kg/ha or Phorate @ 10 kg/ha before one week of transplanting
- Using of biocontrol agents like, *Paecelomyces*, *Metarhizium* and *Beanveria bassiana* and botanical like, Neem cake at the time of transplanting.

# CHAPTER – 6

# Diseases of Malvaceous Crops

## Okra/Ladies finger/Bhindi

### 1. Leaf blight/Spot

#### Pathogen

*Cercospora abelmoschi* Ell. & Ev.

**Symptoms**

The fungus grows as shooty to dark olivaceous, ectoparasite covering entire leaf lamina. Pathogen initiates as small isolated fungal growth on lower side of leaf but very soon it spreads on both sides. Pathogen reduces severely the photosynthesis area. Infected leaves fall down very quickly after dropping and rolling. It causes severe loss in seed production crop as well as in the late sown crops.

#### Management

- Sowing should be completed by first week of July.
- Spray should be started immediately after disease appearance
- Collect the infected leaves and burnt in field to reduce the inoculum.
- Alternate spray of Mancozeb @ 0.25% or Chlorothalonil @ 0.2% at 10-15 days intervals is effective against this disease.

## 2. Wilt of okra

### Pathogen

*Fusarium oxysporum f.sp.vasinfectum* Atk.

### Symptoms

The typical symptom appears as yellowing and stunting of the plant followed by wilting and rolling of the leaves. In later stage the plants are dying. The stem is blackened in case of severe attack.

### Management

- Seed Treatment with Thiram or Captan @ 2.5 g/kg seed or *Trichoderma viride* @ 6-10 g/kg seed before sowing.

**Figure:** Wilt affected okra plant

- Proper drainage is required during entire cropping period.
- Crop rotation with non host crop.
- Use of *Trichoderma viride* @ 4-5 kg/ha in soil before sowing of the corp.

## 3. Powdery Mildew

### Pathogen

*Sphaerotheca fuliginea* Schlecht.

### Symptoms

The fungus produces white greyish powdery growth on the under as well as upper surface of the leaf causing severe reduction in yield. In later stage the leaves turn yellow and finally drop.

**Figure:** Powdery mildew infected okra plant/leaf

### Management

- Spray of sulfex, wettable Sulpher or Karathane @ 0.2% in case of heavy infestation.
- Foliar spray of Penconazole @ 0.05% or Tridemorph @ 0.1% or Carbendazim @ 0.1% is effective
- Infected plant debris would be collected and burn it
- Use of resistant cultivars

## 4. Yellow Vein Mosaic

### Pathogen

Virus (*Gemini virus*)

### Symptoms

Clearing of small veins and veinal chlorosis of leaves. The disease is caused by Gemini virus and transmitted by white fly. The yellow network of veins is very conspicuous and the veins and veinlets are thickened. The chlorosis extends to the interveinal areas and in case of severe infection the whole leaf turns yellow. Fruit develops green to yellow colour and become hard in early stage of growth.

**Figure:** YVMV infected okra plant

### Management

- Alternate spray of any systemic insecticides up to flowering stage.
- Seed treatment with Imidacloprid @ 5 ml/kg of seed.
- Use of resistant cultivars like.VRO-5, VRO-6, Parbhani Kranti, Utkal Gourav and Arka Anamika.
- Early sowing is required in month of June to July.

## 5. Enation Leaf Curl

### Pathogen

Virus (*Gemini virus*)

### Symptoms

The disease is caused by Gemini virus and transmitted by white fly (*Bemisia tabaci*). Symptoms are small green, out growth are observed on vein and vein lets of the lower side of leaves. Stem and leaf petiole are bent. Apical portion of plant develops in to full of small curl and aggregated leaves. A floral part does not set in to the fruit.

## Management

- Seed treatment with Imidacloprid @ 5ml/kg of seed. 2-3 spray of Monocrotophos @ 0.05% up to flowering stage.
- Removal of infected plants at initial stage.
- Use of tolerant/resistant cultivars.

## Root knot Nematode

## Pathogen

*Meloidogyne incognita*

*M. javanica*

## Symptoms

Affected plants are stunted, yellow and have tendency to wilt in hot weather. Very heavily infested plants are killed. Affected plants appear in patches. If infested plants are pulled from the soil, the roots are severely distorted, swollen and have lumps known as galls or root knots. The galls range in size from smaller than a pinhead to 25 mm or more in diameter.

**Figure:** Okra plant severely infested with root knot nematode

### Management

- Soil solarization during nursery raising.
- Crop rotation for 2-3 year with cereal crops.
- Grow resistant cultivars (SL-120, Hissar Lalit etc).
- Using Nemagon @12-15 kg/ha or Phorate @ 10 kg/ha before one week of transplanting
- Using of biocontrol agents like, *Paecelomyces*, *Metarhizium* and *Beaveria bassiana* and botanical like, Neem cake at the time of transplanting

# Annexures

**Table: List of vegetable crops, their diseases and causal organisms**

| S.No. | Name of the Crop | Diseases | Causal agents |
|---|---|---|---|
| 1. | Onion and Garlic | Damping-off | *Pythium aphanidermatum* Edson.) Fitz. *P. debaryanum* *Rhizoctonia solani* Kuehn. *Sclerotium spp* *Fusarium spp* |
| | | Purple blotch | *Alternaria porri* (Ell.) Neerg. |
| | | Stemphylium blight | *Stemphylium vesicarium* Wallr. (Sacc.) |
| | | Cercospora leaf spot | *Cercospora duddiae* Walles. |
| | | Powdery Mildew | *Leveillula taurica* (Lev.) Arnaud |
| | | Downy Mildew | *Pernospora destructor* |
| | | Rust | *Puccinia porri* |
| | | Botrytis leaf blight | *Botrytis ceneria* |
| | | Anthracnose disease (Twister) | *Colletotrichum gloeosporioides* Penz. *Glomerella cingulata* (Stonem) Schrenk & Spauld |
| | | Southern blight | *Sclerotium rolfsii* Sacc. |
| | | White rot | *Selerotium cepivorum* Berk. |
| | | Basal rot/Bottom Rot | *Fusarium oxysporum* f. sp. *cepae* Schlecht. |
| | | Brown rot | *Pseudomonas aeruginosa* |
| | | Soft rot | *Erwinia caratovora* |
| | | Neck rot | *Botrytis allii* Pers. |
| | | Pink Root | *Pyrenochaeta terrestris* de Not. |
| | | Black mould | *Aspergillus niger* van Tiegh. |
| | | Blue mould | *Penicillium corymbiferum*. Link ex Fr. |
| | | Onion smut | *Urocystis cepulae*. Frost. Kolar Dt. |
| | | Sour Skin | *Pseudomonas cepacia* |
| | | Bulb canker/Skin blatch | *Embellissia allii.* |
| | | Garlic Mosaic | Virus |
| | | Iris Yellow Spot Virus (IYSV) | Virus (Tospovirus) |
| | | Onion Yellow Dwarf Virus (OYDV) | Virus (Poty virus) |
| | | Stem and Bulb Nematode | *Ditylenchus dipsaci* |

*(Contd.)*

| S.No. | Name of the Crop | Diseases | Causal agents |
|---|---|---|---|
| 2. | Pea | Rust | *Uromyces fabae* (Pers.) de Bary |
| | | Powdery mildew | *Erysiphe pisi* DC. |
| | | White Rot or Sclerotinia Rot | *Sclerotinia sclerotiorum* de Bary. |
| | | Aschochyta blight | *Aschochyta pisi* Lib. |
| | | Rhizoctonia Seedling Blight | *Rhizoctonia solani* |
| | | Anthracnose | *Colletotrichum pisi* |
| | | Downy Mildew | *Pernospora pisi* (Berk.) de Bary |
| | | Bacterial Blight | *Pseudomonas syringae* pv. *Phaseoli* |
| | | Mosaic | Common mosaic virus |
| 3. | Bean/ Cow Pea | Leaf blight/Spot | *Cercospora cruenta* Sacc. |
| | | Sclerotinia Blight | *Sclerotinia sclerotiorum* (Lib.) de Bary. |
| | | Anthracnose | *Colletotrichum lindemuthianum* Sacc. & Magn. *Glomerella lindemuthianum* |
| | | Downy Mildew | *Peronospora vicae* (Berk.) Gaum. |
| | | Rust | *Uromyces appendiculatus* (Pers.) de Bary |
| | | Bacterial Blight | *Xanthomonas campestris* pv. *Phaseoli Xanthomonas campestris* pv.*phaseoli Pseudomonas syringae* pv. *Phaseolicola* |
| 4. | Tomato | Damping off | *Pythium aphanidermatum* |
| | | Early blight | *Alternaria solani* Ell. & Mart. |
| | | Late blight | *Phytophthora infestans* (de Bary) |
| | | Pseudocercospora Leaf Blight | *Pseudocercospora fuligena* (Lev.) Speg. |
| | | Fruit rots | *Colletotrichum gloeosporoides* Penz. |
| | | Fusarial Wilt in Tomato | *Fusarium oxysporum f. sp. Lycopersici* Brushi. |
| | | Collar rot | *Sclerotium rolfsii* Sacc. |
| | | Sclerotium stem rot | *Sclerotinia sclerotiorum* Lib. de Bary |
| | | Bacterial Wilt | *Pseudomonas solanacearum* |
| | | Bacterial Spot | *Xanthomonas campestris* pv. *vesicatoria* |
| | | Leaf curl | Virus *Gemini Virus* |
| | | Mosaic | Virus *Gemini Virus* |
| | | Root- Knot Nematode | *Meloidogyne incognita M. javanica* |
| 5. | Capsicum | Powdery Mildew | *Leveillula taurica* (Lev.) Arnaud. |
| | | Anthracnose/Fruit rot | *Colletotrichum capsici* syd. Butler & Bisby |
| | | Bacterial Wilt | *Pseudomonas solanacearum* |

*(Contd.)*

| S.No. | Name of the Crop | Diseases | Causal agents |
|---|---|---|---|
| | | Bacterial Spot | *Xanthomonas compestris* pv. *vesicatoria* |
| | | Leaf curl complex | Virus |
| | | Root-Knot of Capsicum | *Meloidogyne hapla* |
| 6. | Chilli | Leaf Spot/Blight | *Alternaria solani* Ell. & Mart. |
| | | Powdery Mildew | *Leveillula taurica* (Lev.) Arnaud. |
| | | Fruit Rot and Die-back | *Colletotrichum capsici* (Syd.) Butler & Bisby |
| | | Leaf Curl | Virus |
| | | White/Sclerotinia Rot | *Sclerotinia sclerotiorum* (Lib.) de Bary |
| 7. | Cole Crops | Leaf spot | *Alternaria brassicae* (Berk.) Sacc. |
| | | Black Leg | *Phoma lingam* (Fr.) Desm. |
| | | Sclerotinia Rot | *Sclerotinia sclerotiorum* (Lib.) de Bary |
| | | Club Root | *Plasmodiophora brassicae* Woron. |
| | | Black Rot | *Xanthomonas campestris* pv *campestris* |
| 8. | Brinjal | Leaf Spot | *Alternaria melongenae* (Rangaswami & sambandam) |
| | | Phomopsis Blight/Fruit Rot | *Phomopsis vexans* (Sacc. & Syd.) Harter. |
| | | Sclerotinia Rot | *Sclerotinia sclerotiorum* (Lib.) de Bary |
| | | Bacterial Wilt | *Ralstonia solacearum* |
| | | Verticillium Wilt | *Verticillium dahliae* (Nees & Wallr.) *Verticillium albo-atrum* (Reink. & Berthhold) |
| | | Little Leaf | Phytoplasma |
| 9. | Cucurbits | Anthracnose | *Colletotrichum laginarium* (Corda.) *Colletotrichum orbaculare* |
| | | Powdery Mildew | *Erysiphe cichoracearum* DC. |
| | | Leaf Spot | *Cercospora citrullina* cooke. |
| | | Downy Mildew | *Pseudopernospora cubensis* |
| | | Wilt of Cucurbits | *Fusarium oxysporum f. sp cucurbitae* |
| | | Mosaic | Virus |
| 10. | Okra | Leaf blight/Spot | *Cercospora abelmoschi* Ell. & Ev. |
| | | Wilt of Okra | *Fusarium oxysporum f.sp.vasinfectum* Atk. |
| | | Powdery Mildew | *Sphaerotheca fuliginea* Schlecht. |
| | | Yellow Vein Mosaic | Virus (Gemini virus) |
| | | Enation Leaf Curl | Virus (Gemini virus) |

**Table:** List of different fungicides/bio agents formulations with different names

| S.N. | Common name | Trade name | Concentration (%) |
|---|---|---|---|
| 1. | Mancozeb | Dithane M-45, Indofil M-45 Dhanuka M-45 | 0.25 |
| 2. | Carbendazim | Bayistin, Dhanustin, Ridomil | 0.1 |
| 3. | Captan | Captan, Dhanutan | 0.2-0.25 |
| 4. | Thiram | Thiaride, Hexaire | 0.2-0.25 |
| 5. | Dinocap | Carathane | 0.2 |
| 6. | Copper oxychloride | Blitox-50, Phytolan, Blue copper, Dhanucop, Tracop | 0.3 |
| 7. | Sulpher | Sulfex, Agrisulf, Dhanusal, Thiovit | 0.2 |
| 8. | Tidemorph | Calixin | 0.2 |
| 9. | Penconazole | Topas | 0.1 |
| 10. | Propineb | Antracol | 0.2 |
| 11. | Chlorothalonil | Kavach | 0.2 |
| 12. | Tebiconazole | Folicure | 0.1 |
| 13. | Diafeniconazole | Score | 0.1 |
| 14. | Hexaconazole | Cantaf, Sitara, Hexzole | 0.1 |
| 15. | Propiconazole | Tilt, | 0.1 |
| 16. | Copper hydraoxide | Kocide | 0.2-0.25 |
| 17. | Mixed fungicides (68% Mancozeb + 12% carbendazim) | Companian, Saaf, Bright, Sixer | 0.2-0.25 |
| 18. | Metalexil + Mancozeb | Ridomil, Metco, Unilex, Egron | 0.2-0.25 |
| 19. | Azoxistrobin | Amister | 0.1 |
| 20. | Bittertanol | Bencor, Veliton | 0.05 |
| 21. | *Trichoderma viride* | Bioderma, Ecoderma, Trichodex | 0.4-0.5 |
| 22. | *Azadirachta indica* | Indoneem, Bioneem,Ozoneem | 0.5 |
| 23. | Furadan | Carbofuran | 3 ac |
| 24. | Carbosufan | Marshal | DS |
| 25. | Nemagon | Nemedyon | |

**Table:** Vegetable cultivars resistant to Annexual diseases

| Crop | Cultivars | Diseases/Insect Pest/Nematodes | Source |
|---|---|---|---|
| Brinjal | BWR-12 | Bacterial Wilt (*Ralstonia solanacearum*) | IIHR-Bangalore |
| | Pant Rituraj | -do- | GBPAUT |
| | Pant Samarat | Bacterial Wilt (*P. solanacearum*) Phomopsis Blight (Phomopsis vexans Shoot & Fruit Borer & Jassids | GBPAUT |
| | Pusa Purple Cluster | Bactetial Wilt (*Ralstonia solanacearum*). | IARI, New Delhi |
| | BB-7 | -do- | OUAT, Bhubaneswar |
| | Pusa Bhairav | Phomopsis Blight (*P. vexans*) | IARI, New Delhi |
| Cabbage | SEL-8 | Black Rot (*Xanthomonas campestris*) | IARI-Katrain |
| Cauliflower | Pua Shubra | Black Rot | IARI, New Delhi |
| | Pusa Snowball K-1 | Black Rot | IARI, Katrain |
| Chilli | Pusa Jawala | Leaf Curl (CMV & PVY) | IARI, New Delhi |
| | Punjab Lal | Leaf Curl (CMV & PVY) | PAU, Ludhiana |
| Cowpea | Pusa Komal | Bacterial Blight (*Xanthomonas vignicola*) | IARI, New Delhi |
| Muskmelon | Arka Rajhans | Powdery Mildew (*Sphaerothica fuliginea*) | IIHR, Bangalore |
| | Punjab Rasila | -do- | PAU, Ludhiana |
| Okra | Sel-10 | Y.V.M Virus | IIHR, Bangalore |
| | Sel-2 | Y.V.M Virus | NBPGR, New Delhi |
| | P-7, P-8 | Y.V.M Virus | PAU, Ludhiana |
| | Parbhani Kranti | Y.V.M Virus | MAU, Prabhani |
| Pea | PRS-4 | Powdery Mildew | GBPAUT, Pantnagar |
| | PM-2 | Powdery Mildew | GBPAUT, Pantnagar |
| | JP-4 | Powdery Mildew | GBPAUT, Pantnagar |
| Tomato | BT-1 | Bacterial Wilt | OUAT, Bhubaneswar |
| | Pant Bahar | Varticilium Wilt & Fusarium Wilt (*Fusarium oxysporum*) | GBPAUT, Pantnagar |
| | SEL-120 | Root Knot Nematode (*M. incognita, M.arneria*) | IARI, New Delhi |
| | PNR-7 | *M. javanica* | CCSHAU, Hisar |
| | Hisar Lalit | RKN | CCSHAU, Hisar |

## List of Pesticides/Pesticides Formulations Banned in India

### A.Pesticides Banned for manufacture, import and use in India (28 Nos.)

1. Aldrin
2. Benzene Hexachloride
3. Calcium Cyanide
4. Chlordane
5. Copper Acetoarsenite
6. CIbromochloropropane
7. Endrin
8. Ethyl Mercury Chloride
9. Ethyl Parathion
10. Heptachlor
11. Menazone
12. Nitrofen
13. Paraquat Dimethyl Sulphate
14. Pentachloro Nitrobenzene
15. Pentachlorophenol
16. Phenyl Mercury Acetate
17. Sodium Methane Arsonate
18. Tetradifon
19. Toxafen
20. Aldicarb
21. Chlorobenzilate
22. Dieldrine
23. Maleic Hydrazide
24. Ethylene Dibromide
25. TCA (Trichloro acetic acid)
26. Metoxuron
27. Chlorofenvinphos
28. Lindane (Banned vide Gazette Notification No S.O. 637(E) Dated 25/03/2011)-Banned for Manufecture,Import or Formulate w.e.f. 25th March,2011 and banned for use w.e.f. 25th March,2013.

### B.Pesticide / Pesticide formulations banned for use but their manufacture is allowed for export (2 Nos.)

29. Nicotin Sulfate
30. Captafol 80% Powder

### C.Pesticide formulations banned for import, manufacture and use (4 Nos)

1. Methomyl 24% L
2. Methomyl 12.5% L
3. Phosphamidon 85% SL
4. Carbofuron 50% SP

### D.Pesticide Withdrawn(7 Nos)

1. Dalapon
2. Ferbam
3. Formothion
4. Nickel Chloride
5. Paradichlorobenzene (PDCB)
6. Simazine
7. Warfarin

*Source*: sDirectorate of Plant Protection, Quarantine and Storage, Faridabad

FSC
www.fsc.org
MIX
Papier | Fördert gute Waldnutzung
FSC® C083411

Zeitfracht Medien GmbH
Ferdinand-Jühlke-Straße 7
99095 Erfurt, Deutschland
produktsicherheit@kolibri360.de